ESSENTIALS ON DARK MATTER

Edited by **Abraão Jessé Capistrano de Souza**

Essentials on Dark Matter

http://dx.doi.org/10.5772/intechopen.72066

Edited by Abraão Jessé Capistrano de Souza

Contributors

Brian Albert Robson, Kevin Ludwick, Andrzej Radosz, Andy T. Augousti, Pawel Gusin, Abraao Jesse J.S. Capistrano

First published in London, United Kingdom, 2018 by IntechOpen

IntechOpen is the global imprint of INTECHOPEN LIMITED, registered in England and Wales, registration number: 11086078, The Shard, 25th floor, 32 London Bridge Street

London, SE19SG – United Kingdom

Printed in Croatia

British Library Cataloguing-in-Publication Data

A catalogue record for this book is available from the British Library

Additional hard copies can be obtained from orders@intechopen.com

Essentials on Dark Matter, Edited by Abraão Jessé Capistrano de Souza

p. cm.

Print ISBN 978-1-78923-680-4

Online ISBN 978-1-78923-681-1

We are IntechOpen,
the world's leading publisher of
Open Access books

Built by scientists, for scientists

3,700+
Open access books available

115,000+
International authors and editors

119M+
Downloads

Our authors are among the

151
Countries delivered to

Top 1%
most cited scientists

12.2%
Contributors from top 500 universities

Interested in publishing with us?
Contact book.department@intechopen.com

Numbers displayed above are based on latest data collected.
For more information visit www.intechopen.com

Meet the editor

Professor Abraão Capistrano graduated with a Physics degree awarded by the Federal University of Pará, Brazil, (2004), masters in Physics (2006), and doctorate in Theoretical Physics from Universidade de Brasília, Brazil (2009). His research is focused on problems in mathematical-physics, quantum fields, astrophysics, cosmology, and gravitation. He also studies rocket science and robotics focused on undergraduate engineering projects as well as teaches physics and astronomy to a broader audience through institutional extension projects. He is currently a full professor at the Federal University of Latin-American Integration and collaborator researcher at Casimiro Montenegro Filho Astronomy Center (Technological Park of Itaipu).

Contents

Introduction

Chapter 1

Introductory Chapter: The Physics of Dark Sector

Abraão Jessé Capistrano de Souza

Additional information is available at the end of the chapter

http://dx.doi.org/10.5772/intechopen.80234

1. On the physics of dark sector

In the last two decades, researches in cosmology and astrophysics provided an important source of data about the gravitational and evolutionary structure of the universe, which stimulates a demand for gravitational theories beyond general relativity in the face of the new conjuncture of the problems of contemporary physics. The physics of dark sector has been one of the most intriguing problems of physics. Since the rise of dark matter problem in the very beginning of the twentieth century and the appearance of the dark energy in the end of 1990s, they launched a new scenario for contemporary physics and some examples of questions can be made such as: Have those problems a true substance of reality? Is there an underlying new physics that describes such issues? Is it possible to explain those problems without changing the ordinary physical theories? Do we need another particle theory? And the central point, what is gravity?

Since the works of Einstein, our sense of gravity had been radically changed since it can be interpreted as a geometrical effect. Einstein's new approach indicated that geometry plays a fundamental role in a physical process. More interestingly, one can state a more profound meaning on gravity as: gravity does not need matter to exist. In Einstein's sense, gravity is also self-interactive, that is, Einstein's field equations exist on vacuum that evinces the grandeur of a geometric approach on a physical theory. In other words, general relativity is the only physical theory that allows a nontrivial vacuum solution. It must be said that it is rather different from Newton's theory of gravitation that a vectoral field of force does exist to provide interaction between two separated masses. Nonetheless, the problem of searching physics for dark sector essentially involves finding eventually new prospects on the meaning of gravity, once dark matter problem and the dark energy are at the first instance effects of gravity.

This book is devoted to discussing fundamental aspects of the dark matter problem. The modern roots on the dark matter problem were basically launched in the 1930s, with Zwicky's observations [1] on a notorious discrepancy of mass in coma cluster that presented 500 times of mass than expected using Newtonian theory (virial theorem). Curiously, this fact passed practically unnoticed in scientific community and it was taken seriously decades later with the observations of galaxies. Only in 1970s, Zwicky's *missing mass problem* was reinforced by Vera Rubin [2] with observations in spiral galaxies showing a huge discrepancy of Newton's law on that scale. Until now, the so-called dark matter problem is still one of the greatest challenges of both observational and theoretical physics.

According to recent observations on Planck collaboration [3], roughly 5% of the universe is known and the rest of it is made of dark components. The dark matter accounts for 26.8% and the dark energy, a sort of energy that may drive the universe to speed up with negative pressure, responds to 68.3% of the universe composition. Moreover, the gravitational field of dark matter cannot be produced by baryons-only by the analysis of the first peak in the power spectrum of the cosmic microwave background radiation. The observations on optical X-ray and gravitational lensing [4] also suggest that the bullet cluster cannot be explained without dark matter.

Since the mid-1980s, astrophysicists have been compiling evidence—such as cosmic microwave background observations, the supernova type Ia data and large-scale structure—that the late-time universe is accelerating. The simplest candidate to explain this acceleration, within the framework of general relativity (GR), is a positive cosmological constant (CC). Many theoretical physicists were reluctant to consider CC as an explanation for acceleration of the universe, since the natural predicted value for CC from particle physics is $\rho_\Lambda \sim 10^{18} \text{GeV}^4$, which has an enormous discrepancy with the astronomical bound for CC, $\rho_\Lambda \sim 10^{-3} \text{ eV}^4$, about 10^{122} times smaller.

The first evidence of a possible accelerated expansion of the universe was obtained through the Hubble Space Telescope of type Ia (SN Ia) supernova in 1998 [5, 6]. The data suggested the existence of some form of energy, or nonagglomerated matter, that should permeate most of the whole universe with negative pressure generating an accelerated expansion of the universe, that is, roughly speaking, providing a repulsive gravitational effect within the scope of general relativity, known as dark energy, and this finding is further reinforced with the agreement of 250 other events in supernova [5, 6] in independent astronomical observations.

One successful theoretical model for explanation of the accelerated expansion is to attribute the dark matter a role in the acceleration problem. The cosmological constant (Λ) plus cold dark matter (CDM) parametrization [7, 8], for short ΛCDM, aims at explaining both formation and growth of large structures in the universe as well as the accelerated expansion problem [9, 10]. One fundamental characteristic that favors the ΛCDM model concerns its applications to cosmological scale and provides a simulation of the growth of the larger structures of the universe consistent with the observations on large scale structure (LSS) surveys [11]. On the other hand, it lacks a more underlying explanation on the nature of the CC itself and dark matter, which leaves unsolved the question from the first principles.

This book is divided into two sections. The chapters aim at discussing different aspects on the dark matter problem by some experts in the field. The first section is devoted to historical aspects of dark matter phenomenology. The author presents a critical review of the dark matter problem. The subsequent chapters discuss technical scientific advances in the field with a study on a mechanism of couplings of dark matter and dark energy. Moreover, a study of black hole physics is present with the research of interior solutions of Schwarzschild black hole, a discussion on black hole thermodynamics and its role in the dark sector. It has been a great opportunity to work again with InTech's editorial team and such an honor to read all the proposed chapters.

Author details

Abraão Jessé Capistrano de Souza

Address all correspondence to: capistranoaj@unb.br

Federal University of Latin-American Integration, Casimiro Montenegro Filho Astronomy Center (Itaipu Technological Park), Foz do Iguassu, PR, Brazil

References

[1] Zwicky F. Die Rotverschiebung von extragalaktischen Nebeln. Helvetica Physica Acta. 1933;**6**:110

[2] Rubin VC, Kent Ford W Jr. Rotation of the Andromeda Nebula from a Spectroscopic Survey of Emission Regions. The Astrophysical Journal. 1970;**159**:379

[3] Ade PAR et al. Planck 2013 results. XVI. Cosmological parameters. Astronomy & Astrophysics. 2014;**571**:A16

[4] Dietrich JP et al. A filament of dark matter between two clusters of galaxies. Nature. 2012;**487**:202

[5] Perlmutter S et al. Measurements of Omega and Lambda from 42 High-Redshift Supernovae. The Astrophysical Journal. 1999;**517**:565

[6] Riess A et al. Observational Evidence from Supernovae for an Accelerating Universe and a Cosmological Constant. Astronomy Journal. 1998;**116**:1009

[7] Padmanabhan T. Cosmological constant-the weight of the vacuum. Physics Reports. 2003;**380**:235

[8] Peebles PJE, Ratra B. The cosmological constant and dark energy. Reviews of Modern Physics. 2003;**75**:559

[9] Spergel DN. The dark side of cosmology: Dark matter and dark energy. Science. 2015; **347**(6226):1100-1102

[10] Arun K, Gudennavar SB, Sivaram C. Dark matter, dark energy, and alternate models: A review. Advances in Space Research. 2007;**60**:166-186

[11] Springel V. The cosmological simulation code gadget-2. Monthly Notices of the Royal Astronomical Society. 2005;**364**:1105

Historical Aspects on Dark Matter Phenomenology

Chapter 2

The Story of Dark Matter

Brian Albert Robson

Additional information is available at the end of the chapter

http://dx.doi.org/10.5772/intechopen.75662

Abstract

Several astronomical observations concerning the structure of galaxies, the rotation of stars in spiral galaxies, the motions of galaxies within a cluster of galaxies, and so on, cannot be understood in terms of Newton's universal law of gravitation and the visible atomic matter within the galactic systems. This chapter reviews the progress made over many decades in the understanding of these cosmological observations that indicate a serious breakdown of Newton's universal law of gravitation unless there exists additional unseen matter, named "dark matter." The only alternative to "dark matter" is to modify Newtonian gravity. The chapter presents a critical review of the two main approaches to providing the additional gravity required to understand the puzzling astronomical observations: (1) the "dark matter" hypothesis providing additional unseen mass and (2) modification of Newton's universal law of gravity such that there is a stronger gravitational field at larger distances. Both Milgrom's modified Newtonian dynamics (MOND) theory and Robson's recent quantum theory of gravity provided by the generation model (GM) of particle physics are discussed.

Keywords: gravity, dark matter, MOND theory, generation model

1. Introduction

The notion of "dark matter" emerged from several astronomical observations concerning the structure of galaxies, the rotation of stars and neutral hydrogen gas in spiral galaxies, the motions of clusters of galaxies, and so on. These observations could not be described in terms of Newton's universal law of gravitation and the visible ordinary atomic matter within the galactic systems. This chapter reviews the progress made over many decades in the understanding of these cosmological observations that indicated a serious breakdown of Newton's universal law of gravitation unless there existed additional unseen matter that was named "dark matter." The only alternative to "dark matter" was to modify Newtonian gravity.

This chapter presents a critical review of the two main approaches to providing the additional gravity required to understand the puzzling astronomical observations: (1) the "dark matter" hypothesis providing additional unseen mass and (2) modification of Newton's universal law of gravity such that there is a stronger gravitational field at larger distances. Both Milgrom's modified Newtonian dynamics (MOND) theory and Robson's recent quantum theory of gravity provided by the generation model (GM) of particle physics are discussed.

2. The notion of dark matter

The notion of "dark matter" emerged from observations of large astronomical objects such as galaxies and clusters of galaxies, which displayed gravitational effects that could not be accounted for by the visible matter: stars, gas, and so on, assuming the validity of Newton's universal law of gravitation.

It was concluded that such observations could only be described satisfactorily if there existed stronger gravitational fields than those provided by the visible matter and Newtonian gravity. Such gravitational fields required either more mass or an appropriate modification of Newton's universal law of gravitation.

Early preliminary evidence for such a "mass discrepancy" was observed in 1933 by Zwicky [1] for the Coma cluster of galaxies. He estimated that the cluster contained considerably more "dark matter" than the visible galactic matter in order to account for the fast motions of the galaxies within the cluster and also to hold the cluster together.

Additional preliminary evidence for the mass discrepancy was found by Babcock [2] in 1939 and Rubin and Ford [3] in 1970 by measuring the *rotation curve* of the Andromeda galaxy, the nearest spiral galaxy to the Milky Way. The rotation curve of a galaxy is the dependence of the orbital velocity of the visible matter in the galaxy on its radial distance from the center of the galaxy. However, neither Babcock nor Rubin and Ford attributed their observations of an increase in mass toward the edge of the galaxy to any missing mass.

In 1970, Freeman [4] found rotation curves for several galaxies that disagreed with expectation based upon the assumption that the galaxies consisted of stars, gas, and nothing else. Freeman suggested that these galaxies, like the Coma cluster observed much earlier by Zwicky, contained considerably more invisible "dark matter" than the luminous matter. In 1973, Roberts and Rots [5], using 21-cm line data, obtained neutral hydrogen rotation curves of three nearby spiral galaxies. These rotation curves extended to considerably larger distances from the centers of the galaxies than the corresponding rotation curves for the stars. In each case, the complete rotation curve was essentially "flat" out to the edge of the 21-cm data.

In 1974, Ostriker et al. [6] stated that the current observed rotation curves strongly indicated that the mass of a spiral galaxy increases approximately linearly with radius to about 1 Mpc so that the ratio of the total mass to the observed visible mass was large. They concluded that the rotation curves could most plausibly be understood if the spiral galaxy was embedded in a giant spherical halo of invisible "dark matter."

Further evidence for the dark matter hypothesis in many spiral galaxies was obtained in the 1970s by Rubin et al. [7], who measured high-quality optical rotation curves for the luminous matter and Bosma [8], who compiled 21-cm rotation curves for the neutral hydrogen gas that extended far beyond the luminous matter of each galaxy. In all these cases, the complete rotation curve was essentially "flat" out to the edge of the 21-cm data.

By 1980, the conclusive observation from the rotation curves of spiral galaxies was that there existed a major "mass discrepancy" that was greater if larger distance scales were involved. This implied that if Newton's universal law of gravitation was approximately valid, as in the Solar System, considerably more mass was required to be present in each galaxy. This invisible matter was termed "dark matter" with the introduction of the *dark matter hypothesis*: each spiral galaxy was embedded in a huge spherical halo of dark matter.

Thus, the notion of "dark matter" essentially emerged from the observed rotation curves of spiral galaxies that provided convincing evidence for a mass discrepancy within the galaxy. The only alternative to "dark matter" seemed to be a significant modification of Newton's universal law of gravitation to provide the required stronger gravitational field at larger distance scales. However, at that time, such a modification of Newtonian gravity was not considered a viable alternative.

3. The dark matter hypothesis

The dark matter hypothesis was essentially established in 1974 by Ostriker et al. [6], who concluded that the rotation curves of spiral galaxies could most plausibly be understood if the spiral galaxy was embedded in a giant spherical halo of invisible "dark matter." In the conventional cosmological model of spiral galaxies [9], each spiral galaxy is considered to be surrounded by a giant halo of invisible (dark) matter that provides a large contribution to the gravitational field at large distances from the center of the galaxy.

Standard model of cosmology [10] assumes that the universe is now composed of about 5% ordinary matter, 27% dark matter, and 68% dark energy, so that dark matter constitutes about 84% of the total mass, while dark energy plus dark matter constitute about 95% of the total mass-energy content of the universe. Thus, for many years, cosmologists have been confronted with the notion that 84% of the gravitational mass of the universe is dark matter.

The hypothesis of a dark matter spherical halo surrounding a spiral galaxy to account for the observed flat rotation curve of the galaxy has yet to be verified. One of the main difficulties is that the nature of the proposed dark matter is *unknown*.

Initially, massive compact halo objects (MACHOs), were searched for within the outer regions of galaxies, using microlensing techniques [11]. The conclusion from these observations was that at most 20% of a galactic halo consists of MACHOs, and the rest of the halo consists of nonbaryonic matter.

The only other known candidates for dark matter are the three neutrinos of the standard model (SM) of particle physics [12]. However, it was demonstrated in 1983 [13] that if dark matter consisted entirely of neutrinos, the large-scale structure of the universe would significantly *differ* from the observed one, since the neutrinos are relativistic particles leading to a smooth large-scale structure. Recently, Frampton [14] has suggested that the nonbaryonic component of dark matter may consist entirely of primordial intermediate mass black holes. However, this suggestion remains to be verified.

The existence of dark matter in the universe suggests that one requires new physics beyond the SM. Three such particles have been searched for without success: (1) axions, (2) weakly interacting massive particles (WIMPS), and (3) sterile neutrinos. These three particles are all *hypothetical* particles, some of which have been introduced into particle physics in order to resolve certain perceived problems.

The axion was postulated in 1977 by Peccei and Quinn [15] in an attempt to understand the strong CP problem in quantum chromodynamics (QCD). To date various experiments have been carried out but *none* have successfully identified an axion particle.

A weakly interacting massive particle (WIMP) is considered to be a new elementary particle, which only interacts via gravity and any other weak force. The basic goal of direct detection of a (WIMP) is to measure the energy deposited when it interacts with nuclei in a detector, transferring energy to nuclei. Such direct-detection experiments need to be carried out deep underground to prevent them being swamped by unwanted noise from cosmic ray particles.

The most favored (WIMP) is the lightest neutral stable particle, the *neutralino*, predicted by the supersymmetric (SUSY) theory of particle physics, which provides a significant relationship between elementary bosons and fermions. This relationship resolves several puzzling problems, including the *hierarchy problem*, for example, the extremely large difference in the strengths of the gravitational and weak interactions $\approx 10^{-36}$. However, to date, no evidence for any SUSY particle has been found either at the large Hadron collider (LHC) in CERN or in the many underground detection laboratories. At the LHC, no previously *unknown* particles, which may be evidence of SUSY, have been observed since the claimed detection of the Higgs boson, so that SUSY probably does not exist. In addition, no (WIMP) has clearly been detected over several decades at any of the underground laboratories such as the large underground xenon (LUX) experiment in the Homestake Mine, Dakota.

However, there has been one claim of direct detection of dark matter from the DAMA-LIBRA experiment at the Gran Sasso laboratory [16]. This experiment has observed a possible dark matter event rate that modulates annually as the Earth travels around the Sun, while the Solar System moves within the disk of the Milky Way and hence through the hypothesized galactic dark matter halo. The count rate is expected to depend upon the relative velocity of the detector and undergoes a modulation that peaks in June, when the relative velocity is at its maximum.

This observation of the DAMA-LIBRA experiment is controversial, since it has been excluded by observations from several direct-detection experiments, including perhaps the most sensitive one, the LUX experiment. In order to test the DAMA-LIBRA claim, a more sensitive direct-detection experiment, SABRE, is being undertaken with improved but similar equipment in Australia [17].

Sterile neutrinos are also hypothetical neutral particles that emerged from the development of the electroweak theory by Glashow [18, 19], who separated the neutrinos into left-handed and right-handed particles. The left-handed neutrinos interact via the left-handed weak interaction, while the right-handed neutrinos do not and only interact via gravity. The right-handed neutrinos correspond to the so-called sterile neutrinos.

The possible existence of sterile neutrinos arose in the development of the SM at a time when the neutrinos were considered to be *massless*. This is no longer the case so that the three "normal" neutrinos are expected to have right-handed components with the same mass as the left-handed components and hence are unsuitable as candidates for dark matter.

4. Milgrom MOND theory

In view of the considerable uncertainties concerning the existence and nature of the proposed dark matter, there have been several attempts to modify Newton's universal law of gravitation instead of introducing dark matter.

In 1983, Milgrom [20] developed a modification of Newtonian dynamics known as the MOND theory, as a possible alternative to dark matter. This theory is based upon describing two astronomical observations: (1) the flat rotational curves of spiral galaxies at large distances from their centers and (2) the Tully-Fisher empirical relation [21], which states that the intrinsic luminosity L (proportional to the total visible mass) of a spiral galaxy and the velocity, v_f, of the matter circulating at the extremities of the galactic disks are given by:

$$L \propto v_f^\alpha, \tag{1}$$

where α is approximately 4.

In order to describe both the flat rotation curves of spiral galaxies and the Tully-Fisher relation, Milgrom suggested that gravity varies from the prediction of Newtonian dynamics for *low* accelerations. In particular, the transition from $1/r^2$ to $1/r$ gravity should occur below a critical "acceleration" a_0 rather than beyond a critical distance r_0: the former leads to the Tully-Fisher relation, while the latter leads to:

$$L \propto v_f^2, \tag{2}$$

in gross disagreement with the Tully-Fisher relation.

The modified law of gravity in terms of a_0 is [22]:

$$g = GM/r^2 + (GMa_0)^{\frac{1}{2}}/r, \tag{3}$$

where the first term corresponds to distances for which the acceleration is $\gg a_0$ and the second term corresponds to distances associated with the flat rotation curves, that is, with v_f. Indeed the second term gives:

$$v_f = (GMa_0)^{\frac{1}{4}}, \tag{4}$$

which, if the mass to luminosity ratio, M/L, is roughly constant for galaxies, leads to the Tully-Fisher relation.

To summarize: MOND is an empirical modification of Newton's gravitational interaction that is designed to provide agreement with two overarching observational facts: (1) the flat rotation curves of spiral galaxies and (2) the Tully-Fisher relation. It achieves this aim by causing the gravitational interaction to change from $1/r^2$ for small distances, r, to $1/r$ at large galactic distances as the gravitational acceleration becomes less than a critical small acceleration $a_0 \approx 1.2 \times 10^{-10}$ m s^{-2}.

However, MOND is *incomplete* in the sense that in order to be more acceptable to the overall scientific community, it needs to be related to a more general underlying theory of gravity. Just as Kepler's laws of planetary motion described mathematically but without any physical content the observed orbits of the planets, it required Newton's universal law of gravitation to understand the physics underlying Kepler's laws.

5. Robson quantum theory of gravity

The generation model (GM) [23] of particle physics has been developed over many years as a viable alternative to the standard model (SM) [12] of particle physics. The SM is considered by the majority of physicists to be *incomplete* in the sense that it provides no understanding of many empirical observations including the existence of three families or generations of leptons and quarks, which apart from mass have similar properties, a *nonunified* description of the origin of mass, and the nature of the gravitational interaction.

The GM overcomes the *incompleteness* inherent in the SM by introducing three important different assumptions [24]: (1) a simplified *unified* classification scheme of the leptons and quarks in terms of additive quantum numbers, (2) an alternative version of *quark mixing* for hadronic processes, and (iii) the weak interactions are *not* fundamental interactions.

The development of the GM, primarily to describe the three generations of leptons and quarks of the SM [25], employing a *unified* classification scheme involving only three *conserved* additive quantum numbers, led to a composite model of the leptons and quarks and also the weak bosons, W and Z, mediating the weak interactions [23, 24].

Thus, the essential difference between the GM and the SM is that in the GM the leptons, quarks, and weak bosons are *composite* particles rather than elementary particles as in the SM.

In the GM, the leptons, quarks, and weak bosons consist of massless spin-1/2 particles called rishons and/or their antiparticles (antirishons). Each rishon carries a single color charge—red, green or blue—and each antirishon carries an anticolor charge—antired, antigreen or antiblue.

The first generation of leptons and quarks comprising the electron, the electron neutrino, the up quark, and the down quark are composed of two kinds of rishons: a T-rishon with electric charge $Q = +\frac{1}{3}$ and a V-rishon with $Q = 0$ and/or their antiparticles: a $\overline{T}$-antirishon with electric charge $Q = -\frac{1}{3}$ and a $\overline{V}$-antirishon with $Q = 0$. Both the T-rishon and the V-rishon were introduced in 1979 by Harari [26] in his schematic model of the first generation of leptons and quarks describing their electric charge states.

The second and third generations of leptons and quarks are composed of the same "core" rishons and/or antirishons as the first generation plus the addition of one and two rishon-antirishon pair(s), Π, respectively, where

$$\Pi = [(\overline{U}V) + (\overline{V}U)]/\sqrt{2}, \tag{5}$$

and the U-rishon has $Q = 0$ and carries a single color charge [23].

The constituents of the leptons and quarks are bound together by a strong QCD color-type interaction [27], corresponding to a local gauged $SU(3)$ field (analogous to QCD in the SM) mediated by massless *hypergluons* (analogous to gluons in the SM).

The nature of the hypergluon fields acting between the rishons and/or antirishons of the composite leptons and quarks are analogous to the gluon fields acting between quarks and/or antiquarks in the SM. In particular, the nature of the hypergluon fields is such that they lead to a runaway growth of the fields surrounding an isolated color charge, implying that an isolated rishon or antirishon would have an infinite energy associated with it [28]. Nature requires such infinities to be essentially canceled or at least made finite. It does this for the composite systems of rishons and/or antirishons by requiring that the composite particle be *colorless*. However, quantum mechanics prevents these color charges from occupying exactly the same place so that the color fields are not exactly canceled although sufficiently to remove the infinities associated with isolated rishons or antirishons.

In the GM, each lepton of the first generation is *colorless* being composed of three antirishons carrying different anticolors. On the other hand, each quark of the first generation is *colored* being composed of one rishon and one colorless rishon-antirishon pair [23].

The second and third generations are identical to the first generation plus one and two *colorless* rishon-antirishon pairs, respectively, so that all leptons are *colorless* and all quarks are *colored*. Consequently, leptons do not combine to form more complex systems, while the quarks form hadrons that consist of two families: *colorless* baryons, made of three quarks with different color charges, and *colorless* mesons, made of one quark and one antiquark with opposite color charges [23].

Within the framework of the GM, the assumption that the elementary particles of the SM—the six leptons, six quarks, and three weak bosons—are all *composite* particles has led to a *unified* origin of mass [29] and a *quantum* theory of gravity [30].

In 1905, Einstein concluded [31] that the mass of a body m is a measure of its energy content E and is given by $m = E/c^2$, where c is the speed of light in a vacuum. Recently, this relationship has been verified [32] to within 0.00004% for atomic systems.

In the SM, the mass of a hadron arises mainly from the energy content of its constituent quarks and gluons, in agreement with Einstein's conclusion. However, the masses of the elementary particles—the leptons, quarks, and weak bosons—are interpreted [33] in a completely different way involving a Higgs field [34, 35]. Thus, the SM does not provide a *unified* origin of mass, contrary to Einstein's conclusion. Furthermore, the so-called Higgs mechanism does *not* provide any physical explanation for the origin of the masses of the leptons, quarks, and weak bosons, as pointed out by Lyre [36].

In the GM, the elementary particles of the SM are *composite* particles. Since the mass of a hadron originates mainly from the energy of its constituents, the GM postulates that the mass of a lepton, quark, or weak boson arises from a characteristic energy associated with its constituent massless rishons, antirishons, and hypergluons. The mass of each of these composite particles arises from the energy stored in the motion of the rishons and/or antirishons and the energy of the color hypergluon fields, E, according to Einstein's equation $m = E/c^2$. Thus, unlike the SM, the GM provides a *unified* description of the origin of all mass and hence has no requirement for a Higgs field to generate the mass of any particle.

Since, to date, there is no direct evidence for any substructure of leptons or quarks, it is expected that the rishons and/or antirishons of each lepton or quark are localized within a very small volume of space by the strong "intrafermion" color interactions, acting between the colored rishons and/or antirishons.

In the GM, the mass hierarchy of the three generations arises from the substructures of the leptons and quarks. The mass of a composite particle will be *greater* if the degree of localization of its constituents is *smaller*, as a consequence of the nature of the strong intrafermion color interactions possessing the property of *asymptotic freedom* [37, 38], whereby the color interactions become stronger for larger separations of the color charges, as a result of *antiscreening* effects. In addition, particles with two or more like electrically charged rishons or antirishons will have larger structures due to electric repulsion. Ref. [23] presents a qualitative understanding of the mass hierarchy of the three generations of leptons and quarks: a quantitative calculation of the mass hierarchy requires very sophisticated computations.

On the other hand, the SM, involving the Higgs field to generate the masses of its elementary leptons and quarks, is unable to provide any understanding of the mass hierarchy of the three generations. As Lyre [36] has pointed out, the introduction of the Higgs field into the SM simply corresponds mathematically to putting in "by hand" the masses of the elementary particles of the SM.

The GM also provides a quantum theory of gravity. Gravitational interaction acts between particles with mass. Such particles are composed of rishons and/or antirishons that carry colored or anticolored charges and hence are required to be *colorless* in order to avoid infinite energies within their systems.

In the GM, the constituent electrons, neutrons, and protons of ordinary matter are all *composite* and *colorless* fermion particles. Between any two such fermion particles, there exists a *residual* interaction arising from the color interactions acting between the rishons and/or antirishons of one fermion and the color-charged constituents of the other fermion. Robson proposed [29, 30] that such "interfermion" color interactions could be identified with the usual gravitational interaction.

In the GM, gravity essentially emerges from the residual color forces between all electrons, neutrons, and protons. This leads [22, 39] to a *new law of gravity*: the residual color interactions between any two bodies of masses m_1 and m_2, separated by a distance r, lead to a universal law of gravitation, which closely resembles Newton's original law given by:

$$F = H(r)m_1m_2/r^2, \tag{6}$$

where Newton's gravitational constant is replaced by a function of r, $H(r)$.

Both the fundamental intrafermion and the residual interfermion color interactions possess two properties arising from the *self-interactions* of the hypergluons mediating these interactions: (1) asymptotic freedom and (2) color confinement [39].

The antiscreening effects arising from the self-interactions of the hypergluons cause the color interactions to become stronger for larger separations of the color charges. In the case of the fundamental intrafermion interactions, this results in an increase in the characteristic energy and hence the mass of a composite particle that is less localized, as discussed earlier. In the case of the residual interfermion (gravitational) interactions acting between two masses, it leads to an increase in the strength of the gravitational interaction for larger separations so that $H(r)$ becomes an increasing function of r.

It is known from particle physics that the strong color interactions tend to increase with the separation of color charges, and for large separations, this increase is approximately a linear function of r [40], in agreement with the flat rotation curves observed for spiral galaxies. Thus, $H(r)$ is expected to be approximately a linear function of r:

$$H(r) = G(1 + kr/r_S), \tag{7}$$

where G is Newton's gravitational constant, k represents the relative strengths of the modified and Newtonian gravitational fields, and r_S is a radial length scale dependent upon the radial mass distribution of the spiral galaxy so that r_S varies from galaxy to galaxy.

Thus, the modified law of gravity in the GM may be written as:

$$g = GM/r^2 + GMk/(rr_S). \tag{8}$$

Eq. (8) is very similar to Eq. (3) of the MOND theory and one can relate the modified terms in the two gravitational acceleration expressions to obtain:

$$a_0 = GM(k/r_S)^2. \tag{9}$$

Thus, the scale length r_S may be regarded as the radial parameter beyond which weak acceleration takes place. Eq. (9) implies that the physical basis of the critical weak acceleration a_0 of the MOND theory is the existence of a radial parameter r_S that defines a region beyond which the gravitational field behaves essentially as $1/r$.

To summarize: gravity in the GM is identified with the very weak, universal, and attractive residual color interactions acting between the particles of ordinary matter. This gravitational interaction is mediated by hypergluons, which self-interact, leading to a significant modification of Newton's universal law of gravitation, especially at galactic distance scales. However, the self-interactions of the hypergluons *cease* at a sufficiently large distance as a consequence of the color confinement property associated with the QCD-like gravitational interaction. This leads to a *finite range* of the gravitational interaction for very large cosmological distances, estimated to be ≈ several billion light years [39].

Eq. (8) describes both the flat rotation curves of spiral galaxies and also the Tully-Fisher relation [22]. This modification of Newton's universal law of gravitation is essentially equivalent to that of the MOND theory, in that both describe these two overarching observational facts. However, the GM is based upon a quantum field theory of gravity, which provides a general underlying theory of gravity and hence a more physical understanding of the MOND results. Furthermore, unlike the MOND theory, the quantum theory of gravity provides a possible understanding of the observed "accelerating" expansion of the universe [41, 42].

6. Conclusion and discussion

In 1980, the conclusive observation from the flat rotation curves of spiral galaxies was that there existed a significant "mass discrepancy" in spiral galaxies, which was greater if larger distance scales were involved. The flat rotation curves indicated that either Newton's universal law of gravitation and hence the General Theory of Relativity [43] required modification at galactic distances or significantly more mass than the visible mass was required to be present within each galaxy.

The possible mass discrepancy within a galaxy led to the *dark matter hypothesis* whereby each spiral galaxy is embedded within a huge spherical halo of dark matter. The only alternative to dark matter was considered to be an appropriate modification of Newtonian gravity to provide the required extra gravitational field at large (galactic) distances.

Initially, the dark matter hypothesis was favored, since, at that time, a significant modification of Newtonian gravity was not considered a viable alternative. However, the dark matter hypothesis has several problems: (1) the *nature* of the proposed dark matter is *unknown*, although it is considered to be mainly nonbaryonic matter, (2) a dark matter halo has not yet been detected *directly*, (3) the density profile of a typical halo is required to be *fine-tuned* in order to produce the observed flat rotation curve of a spiral galaxy, and (4) the lack of dark matter in large globular clusters, which have about the same mass as the smallest dwarf galaxies that are considered to have considerable amounts of dark matter, is a mystery.

Several hypothetical particles have been suggested for the nonbaryonic component of dark matter but, to date, no clear evidence for the existence of any of these particles (axions, WIMPS, or sterile neutrinos) has been obtained.

More recently, modified gravity theories such as the MOND theory have gained popularity, since they overcome most of the problems associated with the dark matter hypothesis. In particular, Mond theory describes the flat rotation curves of spiral galaxies without fine-tuning, and the globular clusters' lack of dark matter is expected to arise from their much smaller size relative to a dwarf galaxy.

The gravitational interaction of the GM, identified with the very weak, universal, and attractive *residual QCD color interactions* acting between ordinary matter particles, is essentially equivalent to that of the MOND theory, in that both describe successfully the flat rotation curves of spiral galaxies and the Tully-Fisher relation. However, the GM quantum theory of gravity, based upon a quantum field theory, provides not only a general underlying theory of gravity and hence a more physical understanding of the MOND theory but also a possible understanding of the so-called *dark energy* causing the observed *accelerating expansion* of the universe. Indeed, the GM quantum theory describes both dark matter and dark energy in terms of two intrinsic properties of the residual QCD color interactions: *antiscreening* at galactic distances and *finite range* at cosmological distances.

The continuing success [44, 45] of the MOND theory together with the underlying GM quantum field theory of gravity is a strong argument against the existence of undetected dark matter halos consisting of unknown matter embedding galaxies.

However, a direct empirical proof of the existence of dark matter is claimed to be provided by two colliding galaxies known as the "bullet cluster" [46]. Observations of the bullet cluster indicate that during the merging process, the dark matter, deduced from gravitational lensing, has passed through the collision point, while the baryonic component of matter, deduced from X-ray emission, has slowed down due to friction and has coalesced within a central region of the combined cluster. This separation of the two kinds of matter is claimed to provide evidence for dark matter. Unfortunately, a similar separation of the regions of non-Newtonian gravity in the MOND and GM gravity theories are expected to occur in the merging of two galaxies such as the bullet cluster, so that these modified gravity theories also describe the merging of the bullet cluster.

Author details

Brian Albert Robson

Address all correspondence to: brian.robson@anu.edu.au

The Australian National University, Canberra, ACT, Australia

References

[1] Zwicky F, Die Rotverschiebung von extragalaktischen Nebeln. Helvetica Physica Acta. 1933;**6**:110-127

[2] Babcock HW. The rotation of the Andromeda nebula. Lick Observatory Bulletin. 1939;**19**: 41-51

[3] Rubin VC, Ford WK. Rotation of the Andromeda nebula from a spectroscopic survey of emission regions. The Astrophysical Journal. 1970;**159**:379-403

[4] Freeman KC. On the disks of spiral and S0 galaxies. The Astrophysical Journal. 1970;**160**: 811-830

[5] Roberts MS, Rots AH. Comparison of rotation curves of different galaxy types. Astronomy and Astrophysics. 1973;**26**:483-485

[6] Ostriker JP, Peebles PJE, Yahil A. The size and masses of galaxies and the mass of the universe. The Astrophysical Journal. 1974;**193**:L1-L4

[7] Rubin VC, Ford KW, Thonnard N. Rotational properties of 21 Sc galaxies with a large range of luminosities and radii, from NGC 4605 (R = 4 kpc) to UGC 2885 (R = 122 kpc). The Astrophysical Journal. 1980;**238**:471-487

[8] Bosma A. The distribution and kinematics of neutral hydrogen in spiral galaxies of various morphological types. Groningen University; 1978;PhD thesis

[9] Ostriker JP, Steinhardt PJ. The observational case for a low-density universe with a non-zero cosmological constant. Nature. 1995;**377**:600-602

[10] Ade PAR et al. (Planck Collaboration). Planck 2013 results. I Overview of products and scientific results. Astronomy and Astrophysics. 2014;**571** Article ID A1, 48pp

[11] Alcock C et al. The MACHO project: Microlensing results from 5.7 years of large magellanic cloud observations. Astrophysical Journal. 2000;**542**:281-307

[12] Gottfried K, Weisskopf VF. Concepts of Particle Physics. Vol. 1. New York: Oxford University Press; 1984. 189p

[13] White SDM, Frenk CS, Davis M. Clustering in a neutrino-dominated universe. Astrophysical Journal. 1983;**274**:L1-L5

[14] Frampton PH. Theory of dark matter. In: Fritzsch H, editor. Cosmology, Gravitational Waves and Particles. Singapore: World Scientific; 2018. pp. 76-89

[15] Peccei RD, Quinn HR. Constraints imposed by CP conservation in presence of pseudoparticles. Physical Review D. 1977;**16**:1791-1797

[16] Belli P et al. Observations of annual modulation in direct detection of relic particles and light neutralinos. Physical Review. 2011;**84**. Article ID 055014 13pp

[17] Froberg F. SABRE: WIMP modulation detection in the northern and southern hemisphere. ArXiv: 1601.05307v1 [physics.ins.det]. 2016. 5pp

[18] Glashow SL. Partial symmetries of weak interactions. Nuclear Physics. 1961;**22**:579-588

[19] Robson BA. The generation model and the electroweak connection. International Journal of Modern Physics E. 2008;**17**:1015-1030

[20] Milgrom M. A modification of the newtonian dynamics as a possible alternative to the hidden mass hypothesis. Astrophysical Journal. 1983;**270**:365-370

[21] Tully RB, Fisher JR. New method of determining distances to galaxies. Astronomy and Astrophysics. 1977;**54**:661-673

[22] Robson BA. The generation model of particle physics and galactic dark matter. International Journal of Modern Physics E. 2013;**22**. Article ID 1350067:11pp

[23] Robson BA. The generation model of particle physics. In: Kennedy E, editor. Particle Physics. Rijeka: InTech; 2012. pp. 1-28

[24] Robson BA. Progressing beyond the standard model. Advances in High Energy Physics. 2013;**2013**. Article ID 341738:12pp

[25] Veltman M. Facts and Mysteries in Elementary Particle Physics. Singapore: World Scientific; 2003. 340p

[26] Harari H. A schematic model of quarks and leptons. Physics Letters B. 1979;**86**:83-86

[27] Halzen F, Martin AD. Quarks and Leptons: An Introductory Course in Modern Particle Physics. New York: Wiley; 1984. 396p

[28] Wilczek F. In search of symmetry lost. Nature. 2005;**433**:239-247

[29] Robson BA. The generation model and the origin of mass. International Journal of Modern Physics E. 2009;**18**:1773-1780

[30] Robson BA. A quantum theory of gravity based on a composite model of leptons and quarks. International Journal of Modern Physics E. 2011;**20**:733-745

[31] Einstein A. Ist die Trägheit eines Körpers von seinem Energieinhalt abhängig. Annalen der Physik. 1905;**18**:639-641

[32] Rainville S et al. World year of physics: a direct test of $E = mc^2$. Nature. 2005;**438**:1096-1097

[33] Aitchison IJR, Hey AJG. Gauge Theories in Particle Physics. Bristol: Hilger; 1982. 341p

[34] Englert F, Brout R. Broken symmetry and the mass of gauge vector bosons. Physical Review Letters. 1964;**13**:321-323

[35] Higgs PW. Broken symmetries and the masses of gauge bosons. Physical Review Letters. 1964;**13**:508-509

[36] Lyre H. Does the Higgs mechanism exist? International Studies in the Philosophy of Science. 2008;**22**:119-133

[37] Gross DJ, Wilczek F. Ultraviolet behavior of non-abelian gauge theories. Physical Review Letters. 1973;**30**:1343-1346

[38] Politzer HD. Reliable perturbative results for strong interactions. Physical Review letters. 1973;**30**:1346-1349

[39] Robson BA. Dark matter, dark energy and gravity. International Journal of Modern Physics E. 2015;**24** Article ID 1550012:10pp

[40] Sumino Y. QCD potential as a "Coulomb-plus-linear" potential. ArXiv: 0303120v3 [hep-ph]. 2003:11pp

[41] Riess AG et al. Observational evidence from supernovae for an accelerating universe and a cosmological constant. Astronomical Journal. 1998;**116**:1009-1038

[42] Perlmutter S et al. Measurements of omega and lambda from 42 high-redshift supernovae. Astrophysical Journal. 1999;**517**:565-586

[43] Einstein A. The basics of general relativity theory. Annalen der Physik. 1916;**49**:769-822

[44] McGaugh SS, de Blok WJG. Testing the hypothesis of modified dynamics with low surface brightness galaxies and other evidence. Astrophysical Journal. 1998;**499**:66-81

[45] Famaey B, McGaugh SS. Modified newtonian dynamics (MOND): Observational phenomenology and relativistic extensions. Living Reviews in Relativity. 2012;**15**. Article ID 10, 159 pp

[46] Clowe D et al. A direct empirical proof of the existence of dark matter. Astrophysical Journal. 2006;**648**:L109-L113

Cosmology and Dark Matter

Chapter 3

Possible Couplings of Dark Matter

Kevin Ludwick

Additional information is available at the end of the chapter

http://dx.doi.org/10.5772/intechopen.77252

Abstract

Dark matter interacts gravitationally, but it presumably interacts weakly through other channels, especially with respect to regular luminous matter. We look at different ways in which dark matter may couple to other fields. We briefly review some example approaches in the literature for modeling the coupling between dark energy and dark matter and examine the possibility of an arguably better-motivated approach via non-minimal coupling between a scalar field and the Ricci scalar, which is necessary for renormalization of the scalar field in curved space-time. We also show an example of a theory beyond the Standard Model in which dark matter is uniquely connected to the inflaton, and we use observational astrophysical constraints to specify an upper bound on the dark matter mass. In turn, this mass constraint implies a limit on the unification scale of the theory, a decoupling scale of the theory, and the number of e-folds of inflation allowed.

Keywords: dark matter, dark energy, inflation, cosmology, astrophysics

1. Introduction

It is fascinating to think that only roughly 4% of our universe is made up of ordinary matter that we are familiar with, while dark matter and dark energy comprise the rest. We still do not understand the fundamental nature of dark matter or dark energy.

Dark matter has only been detected gravitationally so far, and the candidates for dark matter include macroscopic objects, such as black holes and massive compact halo objects (MACHOs), and many non-baryonic particle models [1], including weakly interacting massive particle (WIMP) models. Dark matter was first inferred from the rotation curves of galaxies [2, 3], which seemed to indicate that there must be some unseen mass providing the gravitational potential needed for the orbiting rates of stellar matter near the outer reaches of galaxies to be as high as what was observed. Direct detection experiments that look for direct interaction between dark matter and a target material have strongly constrained the allowed cross section for many interactions due to non-observation [4, 5], and indirect detection may potentially come from the detection of decay products [6, 7], such as neutrinos that the IceCube experiment may detect [8], or cosmic rays accelerated by supernovae that the AMS-02 experiment has studied [9]. There is currently a 3.5-keV radiation signature coming from certain galaxies (and which is noticeably absent in others) that may be explained by interactions with dark matter [10]. For more review of dark matter, consider [11–13].

In the following, we present interesting aspects of some possible dark matter couplings. We examine a connection between dark matter and other fields via non-minimal coupling (i.e., coupling to other fields through the Ricci scalar). After briefly reviewing some parametrizations of coupled dark matter and dark energy in the literature, we explore in detail the coupling between dark energy and dark matter that must be present simply due to space-time curvature by making some reasonable and general assumptions about the dark energy potential and the coupling strength, and we are able to describe the conversion between dark energy and dark matter without ever explicitly specifying a coupling parametrization. Next, we describe a model beyond the Standard Model called the luminogenesis model, which incorporates in a consistent way the inclusion of dark matter and the inflaton, along with other particles beyond the Standard Model. We describe the unique coupling between dark matter and the inflaton in this model, and we use astrophysical constraints to arrive at an upper bound on the dark matter mass, which in turn constrains the unification scale and another scale of the luminogenesis model, along with the number of e-folds of cosmic inflation allowed.

2. Coupled dark matter and dark energy

Consider the action for general relativity in which dark energy is represented by a real scalar field ($c = 1$):

$$S = S_g + S_\phi + S_\xi + S_m = \int d^4x \sqrt{-g} \left[\frac{R}{16\pi G} - \frac{1}{2} g^{\mu\nu} \nabla_\mu \phi \nabla_\nu \phi - V(\phi) - \frac{1}{2} \xi R \phi^2 \right] + S_m, \quad (1)$$

where the first term is the usual contribution to the Einstein tensor (S_g), the second and third terms are the contribution to the scalar field dark energy (S_ϕ), the fourth term allows for non-minimal coupling of the scalar field (S_ξ), and S_m is the action for the rest of the contents of the universe. S_ξ represents the direct interaction between curvature and the scalar field, and it is necessary for the renormalization of a scalar field in a curved background. Minimizing the action with respect to the metric leads to Einstein's equation,

$$R_{\mu\nu} - \frac{1}{2} R g_{\mu\nu} = 8\pi G T_{\mu\nu} \equiv 8\pi G \left(T_{\mu\nu}[\phi] + T_{\mu\nu}[m] \right), \quad (2)$$

where

$$T_{\mu\nu}[m] = -\frac{2}{\sqrt{-g}}\frac{\delta S_m}{\delta g^{\mu\nu}}, \tag{3}$$

$$T_{\mu\nu}[\phi] \equiv -\frac{2}{\sqrt{-g}}\frac{\delta(S_\phi + S_\xi)}{\delta g^{\mu\nu}}. \tag{4}$$

We have included the variation of the interaction term in $T_{\mu\nu}[\phi]$. There are different ways of accounting for S_ξ [14]. Some choose to include the variation of S_ξ instead in the form of an effective gravitational constant G_{eff} that varies with ϕ, but we choose to have a constant G with an altered stress-energy tensor for ϕ. And it follows that

$$\nabla_\mu T^{\mu\nu} = 0. \tag{5}$$

Each component of the contents of the universe is typically modeled as a perfect fluid so that in the fluid's rest frame

$$T_{\mu\nu}[i] = \text{diag}(\rho_i, p_i, p_i, p_i), \tag{6}$$

where i stands for either ϕ or some other content of the universe, ρ_i is its fluid energy density, and p_i is its fluid pressure.

In standard cosmology, the flat Friedmann-Lemaître-Robertson-Walker (FLRW) metric, which describes a homogeneous and isotropic universe, is typically used:

$$ds^2 = -dt^2 + a^2(t)(dx^2 + dy^2 + dz^2). \tag{7}$$

Using this metric, the solutions to Einstein's equations are called the Friedmann equations:

$$H^2 = \frac{8\pi G}{3}\rho, \tag{8}$$

$$\dot{H} + H^2 = -\frac{4\pi G}{3}(\rho + 3p), \tag{9}$$

where $H \equiv \dot{a}/a$ and $\cdot$ represents differentiation with respect to t.

Energy-momentum conservation, Eq. (5), implies

$$\dot{\rho} + 3H(\rho + p) = 0. \tag{10}$$

This equation can also be obtained from Eqs. (8) and (9) and so is not independent of these. Minimizing the action with respect to the field ϕ results in the equation of motion

$$\ddot{\phi} + 3H\dot{\phi} + V'(\phi) + \xi R\phi = 0, \tag{11}$$

where $'$ represents differentiation with respect to ϕ.

In the concordance model of cosmology, each component of the universe is assumed to be separately conserved, that is,

$$\dot{\rho}_i + 3H(\rho_i + p_i) = 0 \tag{12}$$

for all i. In an interacting fluid model, the total fluid is conserved, but not each component separately. If we consider the late universe dominated by dark matter and dark energy, then

$$\rho = \rho_\phi + \rho_m \text{ and } p = p_\phi + p_{m'} \tag{13}$$

and the interaction between the dark matter and dark energy fluids is typically described as

$$\dot{\rho}_\phi + 3H\left(\rho_\phi + p_\phi\right) = -Q, \tag{14}$$

$$\dot{\rho}_m + 3H(\rho_m + p_m) = Q. \tag{15}$$

A sampling of proposals for the interaction term Q are as follows:

$$Q = \beta H \rho_\phi. \tag{16}$$

$$Q = \beta H \rho_{m'} \tag{17}$$

$$Q = \beta H\left(\rho_m + \rho_\phi\right), \tag{18}$$

$$Q = \beta H \rho_\phi \rho_m / \left(\rho_\phi + \rho_m\right), \tag{19}$$

$$Q = -\beta\left(\dot{\rho}_\phi + \dot{\rho}_m\right). \tag{20}$$

The third interaction term listed here has been used as an approach toward solving the coincidence problem. For more details on these models and others see the review [15]. It has also been shown that some amount of interaction between dark energy and dark matter may alleviate tension between local measurements of H_0 from the Hubble Space Telescope and global measurements of H_0 from the Planck Satellite [16].

We are still ignorant of the fundamental nature of dark matter and dark energy, so they very well may interact directly through an interaction term coupling the dark matter and dark energy fields directly, leading to a particular form of Q. At the very least, these fields should interact through the graviton. Even more so, if ξ is non-zero as the renormalizability of a scalar field in a curved background requires, then the form of Q would be according to the term in the Lagrangian $-\frac{1}{2}\xi R\phi^2$. This term is a clear indication of interaction since R depends on H and $\dot{H}$ in the FLRW metric, and R is clearly dependent on the dark matter (and dark energy) fields via the Friedmann equations, Eqs. (8) and (9), since ρ and p can be expressed in terms of the fields, as we will show. And even present in $\sqrt{-g}$ is a dependence on the field content via Einstein's equation, which relates curvature to mass-energy. The relationship here between curvature and mass-energy is fixed if we treat the background as fixed.

2.1. An approach to the coupling between dark matter and dark energy

We now present a clever procedure of studying the coupling between dark matter and dark energy without out directly specifying a potential $V(\phi)$ for dark energy and without specifying a particular parametrization for Q. Using Eq. (4), one obtains [17].

$$T_{\mu\nu}[\phi] = \nabla_\mu\phi\nabla_\nu\phi - \frac{1}{2}g_{\mu\nu}\nabla^\alpha\phi\nabla_\alpha\phi - V(\phi)g_{\mu\nu} + \xi\left(R_{\mu\nu} - \frac{1}{2}Rg_{\mu\nu}\right)\phi^2 + \xi\left(g_{\mu\nu}\phi^2 - \nabla_\mu\nabla_\nu\phi^2\right).$$

(21)

Since

$$T_{00}[\phi] = \rho_\phi \text{ and } T_{ii}[\phi] = p_\phi \text{ for } i = 1, 2, \text{ or } 3, \tag{22}$$

we have

$$\rho_\phi = \frac{1}{2}\dot{\phi}^2 + V(\phi) + 6\xi H\phi\dot{\phi} + 3\xi H^2\phi^2 \tag{23}$$

and

$$p_\phi = \frac{1}{2}(1 - 4\xi)\dot{\phi}^2 - V(\phi) + 2\xi H\phi\dot{\phi} - 2\xi(1 - 6\xi)\dot{H}\phi^2 - 3\xi(1 - 8\xi)H^2\phi^2 + 2\xi\phi V'(\phi). \tag{24}$$

We specify the usual equation-of-state parametrization for dark energy and dark matter,

$$p_\phi = w_\phi\rho_\phi \text{ and } p_m = w_m\rho_m, \tag{25}$$

and we assume pressureless dark matter,

$$w_m = 0. \tag{26}$$

We use the methodology and results of [18] in what follows. Instead of specifying $V(\phi)$, we simply assume that it is changes slowly. This is a good assumption at least around the present cosmological time, for which w_ϕ seems to be fairly constant (and close to -1) [19]. At the very least, a slowly changing potential is certainly consistent with cosmological data, and this approximation serves as a way of allowing for an explicit calculation of w_ϕ and ρ_ϕ that is valid for a variety of choices for $V(\phi)$. Keeping variation small may also help minimize unknown quantum gravity effects [18, 20, 21].

So we assume slow-roll conditions:

$$\frac{1}{V}\frac{dV}{d\phi} \ll 1, \tag{27}$$

$$\frac{1}{V}\frac{d^2V}{d\phi^2} \ll 1. \tag{28}$$

In addition, we assume

$$|w_\phi + 1| \ll 1, \tag{29}$$

meaning that w_ϕ is very close to -1, which accords with cosmological data. We also assume

$$\xi << 1 \tag{30}$$

for simplification, and this assumption is inclusive of the case in which ξ is $1/6$, the conformal coupling value in four dimensions. With these approximations, an analytic expression for w_ϕ can be obtained:

$$
\begin{aligned}
&1 + w_\phi(a) \\
&= \frac{1}{9} \left\{ \left[\frac{\left[1 + \left(\Omega_{\phi 0}^{-1} - 1\right)a^{-3}\right]\left(1 - \Omega_{\phi 0}\right)}{1 + (a^3 - 1)\Omega_{\phi 0}} \right]^{2 - 8\xi/3} \left\{ 6\sqrt{2}z_0\xi B\left(\left[1 + \left(\Omega_{\phi 0}^{-1} - 1\right)a^{-3}\right]^{-1}; \frac{1}{2} - \frac{4\xi}{3}, -1 + \frac{4\xi}{3} \right) \right. \right. \\
&\left. \left. + \left[\sqrt{3}\lambda_0(1 - 2\xi) - 6\sqrt{2}z_0\xi\right]B\left(\left[1 + \left(\Omega_{\phi 0}^{-1} - 1\right)a^{-3}\right]^{-1}; \frac{3}{2} - \frac{4\xi}{3}, -1 + \frac{4\xi}{3} \right) \right\}^2 \right\},
\end{aligned}
\tag{31}
$$

where a 0 subscript denotes the present time ($a_0 = 1$), $\Omega_{\phi 0}$ is the fraction of the present dark energy density $\rho_{\phi 0}$ out of the present total energy density ρ_0, and we have defined

$$z_0 \equiv \sqrt{\frac{4\pi G}{3}}\phi_0 \text{ and } \lambda_0 \equiv -\frac{1}{\sqrt{8\pi G}V}\frac{dV}{d\phi}\bigg|_{\phi = \phi_0}. \tag{32}$$

According to our assumptions, we expect λ_0 to be very small, and cosmological data for $\Omega_{\phi 0}$ implies that z_0 should be very small, so these these λ_0 and z_0 can be chosen appropriately. The function $B(u; a, b)$ used above is the incomplete beta function:

$$B(u; a, b) = \int_0^u t^{a-1}(1 - t)^{b-1}dt. \tag{33}$$

Under the approximations, we can express $\Omega_\phi(a)$ as

$$\Omega_\phi(a) \equiv \rho_\phi/\rho = \left[1 + \left(\Omega_{\phi 0}^{-1} - 1\right)a^{-3}\right]^{-1}. \tag{34}$$

According to the definition of the incomplete beta function, in Eq. (33), $|u|$ is less than 1, and this is true in the case of Eq. (31) since u is equal to $\Omega_\phi(a)$, which is always less than 1. Also, in Eq. (33), z is greater than 0, and this implies in Eq. (31) that ξ is less than 3/8.

In general (no approximation), because the total pressure p is only due to dark energy,

$$w \equiv \frac{p}{\rho} = \frac{p_\phi}{\rho} = w_\phi\Omega_\phi. \tag{35}$$

And using Eq. (10) and

$$\frac{d}{dt} = aH\frac{d}{da},$$

(36)

it can be shown that in general

$$\rho = \rho_0 \, \text{Exp} \left[-\int_1^a \frac{3(1+w)}{a'} \, da' \right].$$

(37)

Now we have what we need to express what Q would be. Eq. (14) tells us

$$-Q = aH\frac{d\rho_\phi}{da} + 3H\rho_\phi(1 + w_\phi),$$

(38)

and we can express this in terms of our expressions for w_ϕ and Ω_ϕ from Eqs. (31) and (34) using H from Eq. (8) and ρ_ϕ from Eq. (34).

As one might expect, for parameters that accord with cosmological data, Q turns out to be very small around the present. In **Figures 1–3**, $\Omega_{\phi 0}$ is 0.69 (in accordance with recent Planck + WP + BAO + JLA data fits from [22]), and the parameters λ_0 and z_0 are appropriately chosen to be small: $\lambda_0 = 0.01$ and $z_0 = 0.01$. **Figure 1** shows how $-Q$ varies with ξ at the present (redshift $z = 0$). We can see that the magnitude of Q increases with increasing ξ, as one would expect from the ξ coupling term in the Lagrangian. Even for the case when ξ is 0, Q is non-zero; although our plots here have been made using approximations, one can think of this coupling as due to, theoretically, the coupling of $\sqrt{-g}$ multiplying the Lagrangian in the field theory or an explicit interaction term in $V(\phi)$ that couples ϕ and the dark matter field directly; either way, we do not expect a large coupling. **Figures 2** and **3** show redshift z on the horizontal axis ($a = \frac{1}{1+z}$), so time increases toward the left in those plots, and $z < 0$ represents the future. For both of these plots, ξ is set to 0.1. One can see how $-Q$ evolves over time in

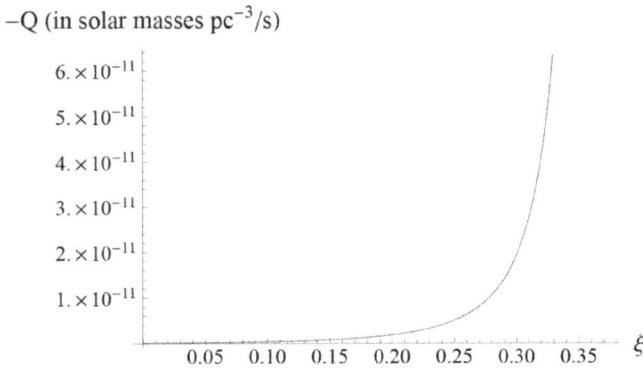

Figure 1. Plot $-Q$ (in solar masses $\times$ parsec^{-3}/second) vs. ξ for the case of redshift $z = 0$, $\Omega_{\phi 0} = 0.69$, $\lambda_0 = 0.01$, and $z_0 = 0.01$.

−Q (in solar masses pc^{-3}/s)

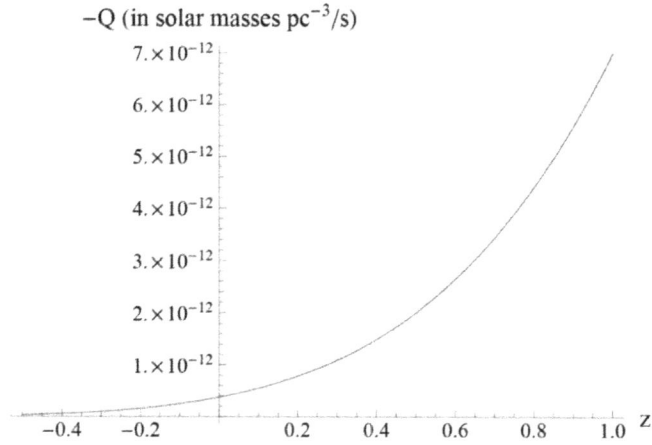

Figure 2. Plot −Q (in solar masses ×parsec^{-3}/second) vs. redshift z for the case of $\xi = 0.1$, $\Omega_{\phi 0} = 0.69$, $\lambda_0 = 0.01$, and $z_0 = 0.01$.

ρ_ϕ,ρ_m (in solar masses pc^{-3})

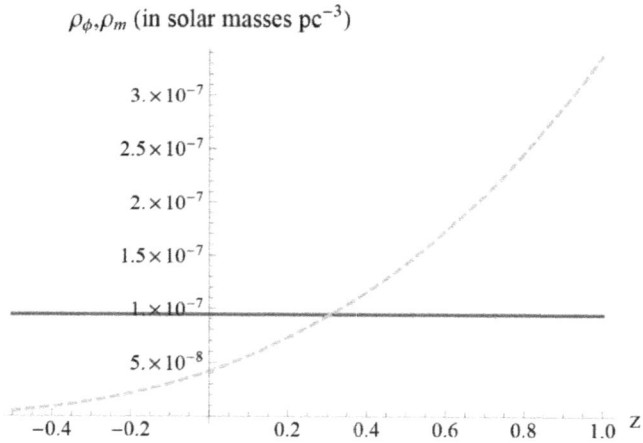

Figure 3. Plots ρ_ϕ and ρ_m (in solar masses ×parsec^{-3}) vs. redshift z for the case of $\xi = 0.1$, $\Omega_{\phi 0} = 0.69$, $\lambda_0 = 0.01$, and $z_0 = 0.01$. ρ_ϕ is represented by the blue solid line, and ρ_m is represented by the dashed green line.

Figure 2. Figure 3 shows how ρ_ϕ acts roughly as a cosmological constant (since we assumed $w_\phi \approx -1$ and strictly bigger than −1) and how ρ_m decreases over time, as expected for cold dark matter.

2.2. How constraints on dark matter may affect inflation

As there is currently no place for a new particle responsible for dark matter in the Standard Model of particle physics, we need a model beyond the Standard Model to include it. One such model is known as the luminogenesis model [23–25]. In the luminogenesis model, dark matter is uniquely connected to the inflaton, as we will discuss, and we are going to utilize astrophysical constraints on strongly-coupled dark matter to constrain its mass, which will allow us to constrain the unification scale and a lower scale of this theory, as well as the number of e-folds of inflation allowed.

The formation of galaxies and galaxy clusters is heavily influenced by the nature of dark matter. For the usual framework of cold dark matter, there are discrepancies between their predictions for them and observations of them. N-body simulations for exclusive collisionless cold dark matter predict the central density profile of dwarf galaxy and galaxy cluster halos to be very cusp-like, whereas observations indicate flat cores (cusp-vs-core problem) [26]. The number of Milky Way satellites predicted in simulations is bigger by an order of magnitude than the number inferred from observations (missing satellite problem) [27, 28], although this may not be very troublesome if more ultra-faint galaxies are successfully detected in the future [29]. The brightest observed dwarf spheroidal galaxy satellites of the Milky Way are predicted to be in the largest Milky Way subhalos, but the largest subhalos are too massive to host them (too-big-to-fail problem) [30]. The resolution of these problems may come through several possible means, including more accurate consideration of baryon interactions, astrophysical uncertainties, and warm dark matter. A promising framework that can solve all these issues is self-interacting dark matter, and that is what we consider in our analysis with the luminogenesis model.

In the luminogenesis model, the dark and luminous sectors are unified above the Dark Unified Theory (DUT) scale. At this DUT scale, the unified symmetry of the model breaks ($SU(3)_C \times SU(6) \times U(1)_Y \rightarrow SU(3)_C \times SU(4)_{DM} \times SU(2)_L \times U(1)_Y \times U(1)_{DM}$), and the breaking is triggered by the inflaton's slipping into the minimum of its symmetry-breaking (Coleman-Weinberg) potential and acquiring the true vacuum expectation value μ_{DUT}, which is the DUT scale energy. This symmetry breaking of $SU(6) \rightarrow SU(4)_{DM} \times SU(2)_L \times U(1)_{DM}$ allows the inflaton to decay to dark matter, and dark matter can in turn decay to Standard Model (SM) and "mirror" matter. The representations of the luminogenesis model (which apply to each of the three families) are given below. The existence of "mirror" fermions, as discussed in [31, 32], is necessary for anomaly cancelation, and it provides a mechanism in which right-handed neutrinos may obtain Majorana masses proportional to the electroweak scale, and they could be searched for at the Large Hadron Collider.

The $SU(4)_{DM}$ dark matter fermions are represented by $(\mathbf{4},\mathbf{1})_3 + (\mathbf{4}^*,\mathbf{1})_{-3}$ in the $\mathbf{20}$ representation of $SU(6)$. The inflaton ϕ_{inf} is represented by $(\mathbf{1},\mathbf{1})_0$ of $\mathbf{35}$, and since $\mathbf{20} \times \mathbf{20} = \mathbf{1}_s +$

$SU(6)$	$SU(4)_{DM} \times SU(2)_L \times U(1)_{DM}$
$\mathbf{6}$	$(\mathbf{1},\mathbf{2})_2 + (\mathbf{4},\mathbf{1})_{-1}$
$\mathbf{20}$	$(\mathbf{4},\mathbf{1})_3 + (\mathbf{4}^*,\mathbf{1})_{-3} + (\mathbf{6},\mathbf{2})_0$
$\mathbf{35}$	$(\mathbf{1},\mathbf{1})_0 + (\mathbf{15},\mathbf{1})_0 + (\mathbf{1},\mathbf{3})_0 + (\mathbf{4},\mathbf{2})_{-3}$
	$+(\mathbf{4}^*,\mathbf{2})_3$

Table 1. $(\mathbf{1},\mathbf{2})_2$ represents luminous matter while $(\mathbf{4},\mathbf{1})_3 + (\mathbf{4}^*,\mathbf{1})_{-3}$ represent dark matter [24, 25].

$35_a + 175_s + 189_a$, the inflaton decays mainly into dark matter χ through the interaction $g_{20} \Psi_{20}^T \sigma_2 \Psi_{20} \phi_{35}$, which contains the inflaton in $g_{20} \chi_L^T \sigma_2 \chi_L^c \phi_{inf}$. The process of luminogenesis refers to the genesis of luminous matter from the initial abundance of dark matter which was formed from the decay of the inflaton. Most indirect detectors of dark matter search for annihilation channels to particle-antiparticle pairs. In the luminogenesis model, dark matter can decay to luminous particle-antiparticle pairs via an effective interaction with the dark photon of $U(1)_{DM}$, but also two χ particles can be converted to a fermion and mirror fermion pair. More details on this model can be found in the aforementioned references.

It is assumed that $(\mathbf{15}, \mathbf{1})_0 + (\mathbf{1}, \mathbf{3})_0 + (\mathbf{4}, \mathbf{2})_{-3} + (\mathbf{4^*}, \mathbf{2})_3$ of $\mathbf{35}$ and $(\mathbf{6}, \mathbf{2})_0$ of $\mathbf{20}$ have masses that are on the order of the DUT scale and thus do not affect the particle theory below that energy scale. Since dark matter should have no $U(1)_Y$ charge, the $SU(4)_{DM}$ particles in $(\mathbf{4}, \mathbf{1})_{-1}$ in the $\mathbf{6}$ representation of $SU(6)$ cannot be dark matter since they have $U(1)_Y$ charge, and they are assumed to decouple below the mass scale we call M_1.

In [25], we make predictions for the mass of χ in the following way:

- We run the $SU(2)_L$ gauge α_2 coupling from the known electroweak scale up to some unknown DUT scale where it intersects with the $SU(4)_{DM}$ gauge coupling α_4.

- Then we run α_4 down to its confinement scale, which is when $\alpha_4 \sim 1$. In analogy with Quantum Chromodynamics (QCD) confinement of $SU(3)_C$, the main contribution to $SU(4)_{DM}$ fermions' dynamical mass is from the confinement scale of $SU(4)_{DM}$, and that energy scale is our dynamical mass prediction for χ.

- In order to specify that scale, we need to specify a DUT scale. Since $SU(6)$ breaks at the DUT scale when the inflaton slips into its true vacuum, we specify the DUT scale and therefore the dynamical mass of χ by constraining the parameters of a symmetry-breaking (Coleman-Weinberg) inflaton potential with Planck's constraints on the scalar spectral index and amplitude.

Using this method and the β-function equation for $SU(4)_{DM}$ and $SU(2)_L$, one can derive a formula for the dynamical dark matter mass m_χ as a function of the DUT scale energy μ_{DUT} and the scale M_1. Assuming M_1 is the only relevant decoupling scale for $SU(4)_{DM}$ below μ_{DUT} and above the known electroweak scale μ_{EW}, we have (from Eq. (10) from [25])

$$m_\chi = \mathrm{Exp}\left[\frac{3\pi}{19} \left(\frac{1}{\alpha_4(\mu_{DM})} - \frac{1}{\alpha_2(\mu_{EW})} \right) \right] M_1^{12/19} \mu_{DUT}^{8/19} \mu_{EW}^{-1/19}, \tag{39}$$

where $\alpha_4(\mu_{DM}) \sim 1$, $\mu_{EW} = 246 \text{ GeV}$, and $\alpha_2(\mu_{EW}) \approx 0.03$. We use this equation to relate μ_{DUT} to M_1 once we have obtained an upper bound on m_χ from astrophysical observational constraints.

Because of the confinement of $SU(4)$, dark baryons are formed from four χ particles. These particles are dubbed CHIMPs, which stands for "χ Massive Particles." A CHIMP is denoted by X, and $X = (\chi\chi\chi\chi)$, and there are three dark flavors of χ, one per luminous family of

QCD. The three flavors enable the CHIMP to have spin zero because its wave function is a product of the $SU(4)$-color singlet wave function, which is antisymmetric, and the spin-space-flavor wave function, which can also be antisymmetric by the appropriate arrangement of 4 χ s, allowing the CHIMP wave function to be symmetric. As we know from QCD, $SU(3)$ Nambu-Goldstone (NG) bosons appearing from the spontaneous breaking of chiral symmetry from $< \bar{q}q > \neq 0$ acquire a small mass from the explicit breaking of quark chiral symmetry due to the small masses of quarks, and they become pseudo-NG bosons known as pions. The small Lagrangian masses of the up and down quarks in QCD (4 and 7 MeV respectively from current algebra) in the terms $m_u \bar{u}u$ and $m_d \bar{d}d$ are much less than their dynamical masses, ~ 300 MeV for both, which is of the order of the QCD confinement scale Λ_3. In QCD, the so-called "constituent masses" of the up and down quarks are for the large part dynamical masses, i.e., $M_{u,d} \sim \Lambda_3$. Also, the pion mass can be obtained from the well-known Gell-Mann-Oakes-Renner relation

$$m_\pi^2 = \frac{m_u + m_d}{2} \cdot \frac{|\langle \bar{q}q \rangle|}{f_\pi^2},$$ (40)

which shows that the pion mass vanishes as $m_u, m_d \to 0$. With $f_\pi \sim \Lambda_3$, it is easy to see that $m_\pi \ll \Lambda_3$. Just as this results from the spontaneous breaking of $SU(3)_L \times SU(3)_R$ in QCD, we expect a similar phenomenon from the condensate $< \bar{\chi}_R \chi_L > \neq 0$ in $SU(4)$, and the NG bosons can acquire a small mass through a term $m_0 \bar{\chi}\chi$ with m_0 a Lagrangian mass parameter for χ which, in analogy with QCD, should obey $m_0 \ll \Lambda_4 \sim m_\chi$. Here m_χ is the *dynamical mass* which is distinct from the *Lagrangian mass* m_0. Similar to what happens in QCD, the dark pion π_{DM} has a mass $m_{\pi_{DM}}$ proportional to m_0 and is expected to be small compared with the dynamical mass m_χ. We seek to constrain the $m_{\pi_{DM}} - m_X$ (m_X being the CHIMP mass) parameter space through astrophysical constraints via the procedure in the following section.

2.3. Solving Schrödinger's equation

For unspecified X and π_{DM}, in general, the cross section of their interaction may not lie in the regimes of the Born or classical approximations, so we cannot rely solely on analytical expressions for these regimes. In order to find how the mass of strongly-coupled DM is correlated to the mass of a scalar mediator via astrophysical constraints, we need to numerically solve Schrödinger's equation, and we use the methodology described in detail in [33].

We take the interaction between dark matter (a CHIMP, denoted by $X = (\chi\chi\chi\chi)$) and a scalar mediator (π_{DM}) to be given by an attractive Yukawa-type potential

$$V(r) = -\frac{\alpha_{DM}}{r} e^{-m_{\pi_{DM}} r},$$ (41)

where $m_{\pi_{DM}}$ is the mass parameter for π_{DM} and the $X - \pi_{DM}$ coupling α_{DM} is represented by the effective interaction

$$\mathcal{L}_{int} = g_{DM}\bar{\chi}\chi\pi_{DM} \tag{42}$$

where α_{DM} is defined as $g^2_{DM}/(4\pi)$. The interaction between the CHIMPs and π_{DM} is via the effective interaction between the scalar and the constituent χ s in Eq. (42), in analogy with the chiral quark model where the gluon fields have been integrated out. Another possibility is to write an effective CHIMP-dark pion interaction Lagrangian, but then the coupling would be dimensionful. We expect g_{DM} to be at least 1 or bigger, and since the pion-nucleon coupling in QCD is $O(10)$, we analyze the cases $\alpha_{DM} = 1$ and $\alpha_{DM} = 10$.

We carried out the computational method for solving Schrödinger's equation exactly as described in [33] with a similar level of computational accuracy for most of the steps, and we plot m_χ vs. $m_{\pi_{DM}}$ for $\alpha_{DM} = 1$ and $\alpha_{DM} = 10$ via their relationship through the velocity-averaged transfer cross section $<\sigma_T>$ for the interaction described by the potential in Eq. (41). The plots are shown in **Figures 4** and **5**.

Using the convention of [33], the plots are described as follows:

- Blue lines going from left to right respectively represent $\langle\sigma_T\rangle/m_\chi = 10$ and 0.1 cm^2/g on dwarf scales, required for solving small scale structure anomalies.

- Red lines going from left to right respectively represent $\langle\sigma_T\rangle/m_\chi = 1$ and 0.1 cm^2/g on Milky Way (MW) scales.

- Green lines going from left to right respectively represent $\langle\sigma_T\rangle/m_\chi = 1$ and 0.1 cm^2/g on cluster scales.

The above astrophysical upper and lower bounds on $\langle\sigma_T\rangle/m_\chi$ are discussed in [33]. They come largely from N-body structure formation simulations for a limited number of specific cross sections, so their constraining power in our plot should not be taken to be extremely stringent.

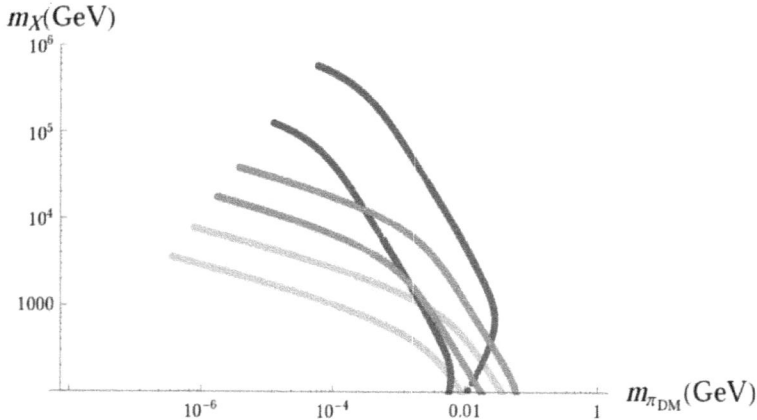

Figure 4. Plot m_χ vs. $m_{\pi_{DM}}$ for the case of $\alpha_{DM} = 1$. We see that all three constraints from clusters (green), the milky way (red), and dwarf galaxies (blue) (described in the text) can be met for m_χ ranging from a few 100 GeV to about 1 TeV since this parameter space falls within all three sets of colored lines.

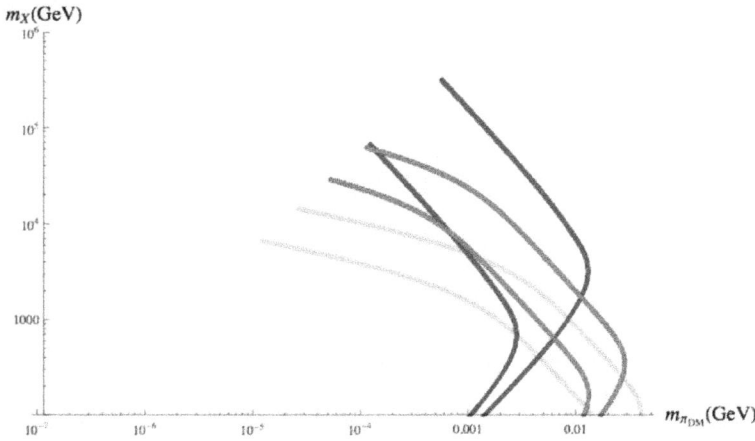

Figure 5. Plot m_X vs. $m_{\pi_{DM}}$ for the case of $\alpha_{DM} = 10$. We see that all three constraints from clusters (green), the milky way (red), and dwarf galaxies (blue) (described in the text) can be met for a range of m_X with an upper limit of about 4 TeV.

But the ranges given for $\langle \sigma_T \rangle / m_X$ are generally what is needed to satisfy observational constraints from structure formation, and we discuss the regions of $m_X - m_{\pi_{DM}}$ parameter space that fall within all three ranges (within the bounds of all three sets of colored lines) of $\langle \sigma_T \rangle / m_X$.

2.4. Analysis of results

We plot the results of our analysis in **Figure 5** for $m_X \gtrsim 100$ GeV. We are primarily interested in this mass range, and this is also the range we examined in [25]. As one can see from **Figure 6** in [33], the resonances present in the three sets of constraints (blue, red, and green lines) become more aligned and overlapped as the coupling parameter α increases. We focused our computing power on calculating data points for $m_X \gtrsim 100$ GeV since we were looking for an upper bound of mass beyond which the three sets of lines do not overlap (i.e., where all three observational constraints are not met). For $1 \lesssim \alpha_{DM} \lesssim 10$, we can see from **Figures 4** and **5** that all constraints from clusters, the Milky Way, and dwarf galaxies can be met for m_X ranging from a few 100 GeV (lower bound from the $\alpha_{DM} = 1$ plot) to about 4 TeV (upper bound from

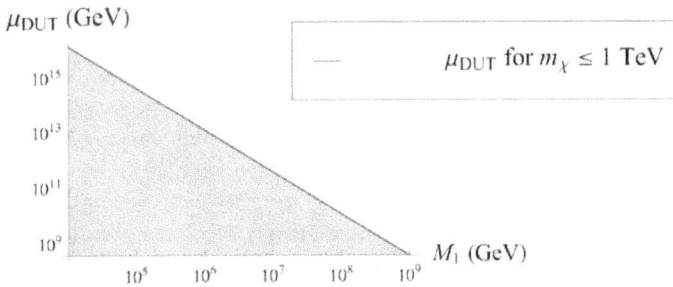

Figure 6. Plot μ_{DUT} vs. M_1 for $m_\chi \lesssim 1$ TeV.

the $\alpha_{DM} = 10$ plot), and this range corresponds to 1 MeV $\lesssim m_{\pi_{DM}} \lesssim 10$ MeV. We point out the noteworthy observation that $m_X \gtrsim 10$ TeV does not agree with all three constraints in the plots (barring the fact that the tightness of these astrophysical constraints is open to interpretation, as discussed in the previous section).

Given the numerical results in the previous paragraph, and since $\Lambda_4 \sim m_\chi \lesssim m_X/4$, one can see from the plots that the approximation $m_{\pi_{DM}} \ll \Lambda_4$ seems to be a good one, much better than the analogous chiral approximation in QCD. This connection between the constraints on the macroscopic astrophysical scale and the microscopic $\pi_{DM} - X$ interaction lends support to the viability of the luminogenesis model.

We now consider the implications of this upper bound on the mass of strongly-coupled dark matter for the luminogenesis model. Since we saw that $X = (\chi\chi\chi\chi)$ cannot have a mass bigger than about 4 TeV, and since $m_\chi \lesssim m_X/4$, we see there is an upper bound of about 1 TeV for m_χ. In **Figure 6**, we plot μ_{DUT} vs. M_1 for this constraint $m_\chi \lesssim 1$ TeV using Eq. (39). From **Figure 6**, we see that $\mu_{DUT} \lesssim 10^{16}$ GeV in order for this astrophysical upper bound for m_χ to be satisfied, and most of the viable parameter space (the shaded triangle) is for values of μ_{DUT} much less than 10^{16} GeV. Along with this constraint, we also see that $M_1 \lesssim 10^9$ GeV to allow for $M_1 \lesssim \mu_{DUT}$.

Using this upper bound on μ_{DUT} along with Planck's constraints on the scalar spectral index and amplitude, we can also determine upper bounds on the number of e-folds and the parameters of the potential for inflation (in our case, the Coleman-Weinberg potential we used in [25]). We work out the relationships of these parameters under the constraints from Planck in Eq. (21) of [25], and one can see that the number of e-folds would need to be less than roughly 95.

3. Conclusion

In general, dark matter is weakly coupled to standard luminous matter (except for gravitational coupling on large scales). However, it is unknown how exactly dark matter interacts with non-standard entities, such as dark energy and the inflaton. We have examined two cases of dark matter coupling.

In the first case, we studied the coupling of dark matter to dark energy without assuming a particular functional form for the conversion rate, and we assumed that dark matter and dark energy were the only components present in the universe. We illustrated a useful way of having interaction between dark matter and dark energy that avoided the need to specify a parametrization for Q, and this is convenient since we do not know what Q should be from first principles. We accomplished our goal by assuming a slowly varying dark energy field and a value of ξ that is very small. We pointed out that, at the very least, there should be coupling between dark matter and dark energy via the ξ term in the Lagrangian necessary for the renormalization of the scalar field for dark energy in a curved background, and we showed in our plots that the magnitude of the coupling Q indeed grew as the coupling constant ξ increased. Of course, one may consider the case of scalar field dark matter, and then another term coupling this dark matter field to R would be present and would indirectly represent another coupling of dark matter and dark energy. Ideally, what is needed is a direct calculation of the cross section between dark energy and dark matter in curved space-time in order to see fundamentally how this non-minimal coupling term affects their interaction. Also, a more accurate treatment would allow for other components of the universe to be present, which would allow for coupling between dark matter and regular luminous matter strictly through curvature via the Ricci scalar, although we would also expect this interaction to be small in general. A more accurate treatment would also allow for back-reaction on the metric and a quantum treatment of gravity itself.

In the second case of dark matter coupling, we showed one way that dark matter may be coupled to the inflaton. We showed an interesting connection between the two fields in the luminogenesis model, which is a unified field theory that consistently combines the Standard Model with other groups that contain dark matter, the inflaton, and other non-standard fields. Using constraints from N-body structure formation simulations, we constrained the mass of self-interacting dark matter, which in turn constrained the DUT scale and the M_1 scale of the luminogenesis model. This constraint on the DUT scale then provided an upper limit on the number of e-folds of inflation allowed in the model.

There are many potential ways in which dark matter couples to other fields, and we simply pointed out interesting facets of two different possible couplings. The true nature of dark matter and how it interacts with other matter is yet to be fully unraveled, but we must pursue every feasible avenue in order to be ready when more precise measurements are available.

Author details

Kevin Ludwick

Address all correspondence to: kludwick@lagrange.edu

Department of Chemistry and Physics, LaGrange College, LaGrange, GA, USA

References

[1] Peebles PJE. Growth of the nonbaryonic dark matter theory. Nature Astronomy. 2017;**1**: 0057

[2] Zwicky F. Die rotverschiebung von extragalaktischen nebeln (In German). The redshift of extragalactic nebulae. Helvetica Physica Acta. 1933;**6**:110

[3] Rubin VC. One Hundred years of rotating galaxies. Publications of the Astronomical Society of the Pacific. 2000;**112**:747

[4] Buchmueller O, Doglioni C, Wang L-T. Search for dark matter at colliders. Nature Physics. 2017;**13**:217

[5]   Liu J, Chen X, Ji X. Current status of direct dark matter detection experiments. Nature Physics. 2017;**13**:212

[6]   Conrad J, Reimer O. Indirect dark matter searches in gamma and cosmic rays. Nature Physics. 2017;**13**:224

[7]   Halzen F. High-energy neutrino astrophysics. Nature Physics. 2017;**13**:232

[8]   IceCube Collaboration, Aartsen MG, et al. Measurement of atmospheric neutrino oscillations at 6–56 GeV with iceCube deepCore. Physical Review Letters. 2018;**120**:071801 arXiv:1707.07081 [hep-ex]

[9]   AMS Collaboration, Aguilar M, et al. Antiproton flux, antiproton-to-proton flux ratio, and properties of elementary particle fluxes in primary cosmic rays measured with the alpha magnetic spectrometer on the international space station. Physical Review Letters. 2016; **117**:091103

[10] Riemer-Sørensen S. Constraints on the presence of a 3.5 keV dark matter emission line from Chandra observations of the Galactic centre. Astronomy and Astrophysics. 2016;**590**:A71 arXiv:1405.7943 [astro-ph.CO]

[11] Bertone G, Hooper D. A history of dark matter. `arXiv:1605.04909 [astro-ph.CO]`

[12] Peter AHG. Dark matter: A brief review. arXiv:1201.3942 [astro-ph.CO]

[13] Roberts MS. M31 and a brief history of dark matter. Astronomical Society of the Pacific. 2008;**3**(**95**):283

[14] Faraoni V. Inflation and quintessence with nonminimal coupling. Physical Review D. 2000;**62**:023504 arXiv:gr-qc/0002091

[15] Wang B, Abdalla E, Atrio-Barandela F, Pavon D. Dark matter and dark energy interactions: Theoretical challenges, cosmological implications and observational signatures. Reports on Progress in Physics. 2016;**79**(9):096901 arXiv:1603.08299[astro-ph.CO]

[16] Yan Y-J, Deng W, Meng X-H. Can decaying vacuum elucidate the late-time dynamics of the universe? arXiv:1801.00689[astro-ph.CO]

[17] Hrycyna O. What ξ? Cosmological constraints on the non-minimal coupling constant. Physics Letters B. 2017;**768**:218-227 arXiv:1511.08736[astro-ph.CO]

[18] Gupta G, Saridakis EN, Sen AA. Non-minimal quintessence and phantom with nearly flat potentials. Physical Review D. 2009;**79**:123013 arXiv:0905.2348[astro-ph.CO]

[19] Frampton PH, Ludwick KJ, Scherrer RJ. The little rip. Physical Review D. 2011;**84**:063003 arXiv:1106.4996[astro-ph.CO]

[20] Huang Q-G. Theoretic limits on the equation of state parameter of quintessence. Physical Review D. 2008;**77**:103518 arXiv:0708.2760[astro-ph]

[21] Saridakis EN. Theoretical limits on the equation-of-state parameter of phantom cosmology. Physics Letters B. 2009;**676**:7 arXiv:0811.1333[hep-th]

[22] SDSS Collaboration, Betoule M, et al. Improved cosmological constraints from a joint analysis of the SDSS-II and SNLS supernova samples. Astronomy and Astrophysics. 2014;**568**:A22 arXiv:1401.4064[astro-ph.CO]

[23] Frampton PH, Hung PQ. Positron excess, luminous-dark matter unification and family structure. Physics Letters B. 2009;**675**:411 arXiv:0903.0358[hep-ph]

[24] Frampton PH, Hung PQ. Luminogenesis from inflationary dark matter. arXiv:1309.17 23[hep-ph]

[25] Hung PQ, Ludwick KJ. Constraining inflationary dark matter in the luminogenesis model. Journal of Cosmology and Astroparticle Physics. 2015;**09**:031 arXiv:1411.1731[hep-ph]

[26] de Blok WGJ. The core-cusp problem. Advances in Astronomy. 2010;**2010**:789293 arXiv: 0910.3538[astro-ph.CO]

[27] Moore B et al. Dark matter substructure within galactic halos. The Astrophysical Journal. 1999;**524**:L19 arXiv:astro-ph/9907411

[28] Klypin AA et al. Where are the missing galactic satellites? The Astrophysical Journal. 1999;**522**:82 arXiv:astro-ph/9901240

[29] Simon JD, Geha M. The kinematics of the ultra-faint milky way satellites: Solving the missing satellite problem. The Astrophysical Journal. 2007;**670**:313 arXiv:0706.0516[astro-ph]

[30] Boylan-Kolchin M, Bullock JS, Kaplinghat M. Too big to fail? The puzzling darkness of massive milky way subhaloes. Monthly Notices of the Royal Astronomical Society. 2011; **415**:L40 1103.0007[astro-ph.CO]

[31] Hung PQ. A Model of electroweak-scale right-handed neutrino mass. Physics Letters B. 2007;**649**:275 arXiv:hep-ph/0612004

[32] Hoang V, Hung PQ, Kamat A. Electroweak precision constraints on the electroweak-scale right-handed neutrino model. Nuclear Physics B. 2013;**10**:002 arXiv:1303.0428[hep-ph]

[33] Tulin S, Yu H-B, Zurek KM. Beyond collisionless dark matter: Particle physics dynamics for dark matter halo structure. Physical Review D. 2013;**87**:115007 arXiv:1302.3898[hep-ph]

Chapter 4

Black but Not Dark

Andrzej Radosz, Andy T. Augousti and Pawel Gusin

Additional information is available at the end of the chapter

http://dx.doi.org/10.5772/intechopen.77963

Abstract

Large black holes of millions of solar masses are known to be present in the centre of galaxies. Their mass is negligible compared to the mass of the luminous matter, but their entropy far exceeds the entropy of the latter by 10 orders of magnitude. Strong gravitational fields make them 'black'—but at the same time, they cause them to emit radiation—so they are not 'dark'. What is the meaning of their borders that may only be crossed once and that leads to the information paradox and what are the properties of their interiors? In discussing these and related questions (is it possible that the volume of a black hole might be infinite?), we uncover the unexpected meaning of the term 'strong gravity'.

Keywords: gravity, black holes, horizon, interior, information paradox

1. Introduction

Black holes (BHs) are sources of the strongest gravitational fields in the Universe. On the other hand, they are also the outcomes of these strong gravitational fields. The first time they appeared in science was as a result of speculation. At the end of the XVIIIC, the English geologist (and astronomer) John Michell and the famous French mathematician Pierre-Simon Laplace independently considered the consequences of the presence of a large, compact massive object producing gravitational fields so strong that even light could not escape from them. For obvious reasons, discussions of this kind were limited in their nature at that time.

The next step came at the beginning of 1916, when Karl Schwarzschild, a mathematician and an army officer, found a specific solution for the field equations of Einstein's General Theory of Relativity. He found the solution for the particular case of a static, spherically symmetric spacetime. Schwarzschild sent the results of these studies to Albert Einstein in the form of two chapters. The second of these two chapters contained what was, to Einstein, a controversial result. If the mass of the source of the gravitational field was both big enough and compact enough, then the solution was singular: a particular element of the metric tensor, a tool for describing the geometric properties of the spacetime, became infinite at some distance from the centre. Einstein was concerned by this effect and consequently had been slow to respond; in the meantime, Schwarzschild had died.

Schwarzschild's solution [1] (see subsequent text) reveals a specific form of behaviour and leads to the conclusion that in some circumstances, a so-called horizon (termed an event horizon) is formed around the black hole. Such a horizon acts as a semi-permeable 'membrane' [2]: it may be crossed only once and in one direction only. The radius of the event horizon is called the gravitational radius or the critical or Schwarzschild radius.

The term 'Black Hole', proposed in the 1960s by J.A. Wheeler, represents the reality of a strong gravitational field in which neither massive nor massless objects (i.e. light in the form of photons) could leave its interior. Black holes (BHs) had been regarded as hypothetical objects even as late as the early 1970s; at that time, a famous bet between two prominent physicists, Kip Thorn (Nobel Prize winner in Physics in 2017) and Stephen Hawking, was set. The subject of the bet was the experimental confirmation of the presence of black holes (the annual delivery of a journal from a building sector was the pledge for this bet).

Currently, it is assumed that there is a massive BH with a mass of millions of solar masses ($M_\odot$) in the centre of each large galaxy [3]. The black hole closest to the Solar System is located at a distance of 1700 light years from us. In the centre of Milky Way, there is a BH of mass $4.3 \cdot 10^6 M_\odot$; one of the largest BHs with a mass of a billion solar masses has been found in the centre of the Sombrero galaxy. This allows us to estimate that the matter confined within black holes is many orders of magnitude smaller than the luminous matter (LM) in each galaxy,

$$\rho_{BH} \leq 10^{-3} \rho_{LM} \tag{1}$$

Hence contributes a negligible fraction of the total energy density. An interesting fact, however, is that the total entropy of black holes, $S_{BH}(tot)$ is 10 s of orders of magnitude higher than the entropy of radiation (CMB), estimated at a value of 10^{90}. Indeed, the entropy of a BH of mass $4.3 \cdot 10^6 M_\odot$ is

$$S_{BH}\left(4.3 \cdot 10^6 M_\odot\right) \cong 10^{90} \tag{2}$$

(see subsequent text), so

$$S_{BH}(tot) \geq 10^{101}, \tag{3}$$

some 20 orders of magnitude smaller than the maximal entropy of our Universe.

The purpose of this exposition is to illuminate the properties of strong gravitational fields. This will be achieved via a discussion of particular processes and phenomena in the vicinity of the event horizon of black holes, on both sides of this horizon.

2. The Schwarzschild solution and the event horizon

Let us consider the case of mass M as the source of a static and isotropic gravitational field. Then, the geometric properties of the resulting spacetime are determined by the Schwarzschild solution, a metric tensor $g_{\alpha\beta}$. The line element, given in terms of Schwarzschild coordinates, $\{x^\alpha\} = t, r, \theta, \varphi$, is (see [1])

$$ds^2 = g_{\alpha\beta}dx^\alpha dx^\beta = f(r)c^2 dt^2 - \frac{1}{f(r)}dr^2 - r^2 d\Omega^2 \tag{4}$$

where $f(r) = 1 - \frac{r_g}{r}$, $r_g = \frac{2GM}{c^2}$ denotes the gravitational radius and $d\Omega^2 = d\theta^2 + \sin^2\theta d\varphi^2$ is a surface element of a unit sphere (we will utilize the system of units such that $c = G = 1$). Solution (2.1) is determined in an empty space outside mass M. Usually, when one deals with a weak gravitational field, the radius R_M of mass M is much larger than its critical radius, $R_M \gg r_g$, then $f(r) \cong 1$. Actually, for the Earth, $r_g(E) \approx 6$ mm, the strength of the gravitational field is of the order of 10^{-9}; the strength of the solar gravitational field is still very weak, 10^{-6}; but neutron stars yield strong gravitational fields, 10^{-1}. Black holes are the sources of the strongest fields, where an event horizon (defined by $f(r) = 0$) is developed. In such a case, we shall consider that the space outside and inside the horizon is empty—the mass of the black hole is confined at $r = 0$—a singularity. This case will be referred to as an eternal black hole. We shall call the region outside the horizon as the exterior and that inside the horizon as the interior of the black hole.

3. Exterior of the Schwarzschild BH

The meaning of a strong gravitational field is revealed via investigation of the properties of the exterior and then the interior of a BH. It is natural to start from the former region. Let us underline the first, nearly trivial fact that the (relativistic) definition of the gravitational radius as the singularity of the metric (2.1) coincides with a purely classical physics definition of a critical radius such that the escape velocity becomes equal to the speed of light in a vacuum, c (see [4]). The generalization of this observation [4, 5] leads to the conclusion that the speed of a freely falling test particle tends to c, independently of the initial conditions. This and the other properties of the exterior of the event horizon may be described by means of geodesics of both kinds, that is, for massive and massless particles (light rays). The geodesic equations may be derived from the following Lagrangian (see Eq. (4)):

$$\mathcal{L} = f(r)\dot{t}^2 - \frac{1}{f(r)}\dot{r}^2 - r^2\dot{\theta}^2 - r^2\sin^2\theta\dot{\varphi}^2 \tag{5}$$

in a standard manner leading to the Euler–Lagrange equations; $\dot{x}^\mu \equiv \frac{dx^\mu}{d\sigma}$ and σ is an auxiliary parameter. There are two conserved quantities resulting from the symmetry conditions: energy, e (due to time independence of the Lagrangian), and angular momentum, l (due to the invariance of the Lagrangian with φ). The latter condition results in the planar character of geodesic motions, so one may, without loss of generality, choose an equatorial plane, $\theta = \frac{\pi}{2}$ and express these conservation laws as follows:

$$f(r)\dot{t} = e, \tag{6}$$

$$r^2\dot{\varphi} = l. \tag{7}$$

One can determine then arbitrary geodesics from the normalization condition

$$g_{\mu\nu}\dot{x}^\mu\dot{x}^\nu = \eta \tag{8}$$

where $\eta = 1$ or 0 for time-like (massive object) geodesics or for light-like (massless object) geodesics, respectively. Indeed, the radial component of the velocity vector, u ($\eta = 1$), or the wave vector, k ($\eta = 0$), takes the form:

$$\dot{r} = \pm\sqrt{e^2 - f(r)\left(\frac{l^2}{r^2} + \eta\right)} \tag{9}$$

Using Eqs. (6)–(9), one can characterize both types of geodesics and illustrate in this way selected features of gravitational fields outside the BH horizon.

Apart from geodesic motions, we will also be employing systems of static observers, SO, whose spatial coordinates are fixed. They are characterized by velocity four-vector,

$$u_{SO} = \left(\frac{1}{\sqrt{f(r)}}, 0, 0, 0\right) \tag{10}$$

3.1. Travel time towards BH horizon

Let us consider the situation of observer A (Alice) whose frame of reference is in a radial free fall, $l = 0$ towards the BH horizon (4). A's frame (or "spaceship") initially was at rest at a Mother Station, MS, located at r_0. The coordinate time to cover the radial coordinate range (r_0, r) in this case is found from Eqs. (6)–(9)

$$t = -\int_{r_0}^{r} \frac{erdr}{\sqrt{(r - r_g)\left[r_g - r(e^2 - 1)\right]}} \tag{11}$$

It diverges, $t \to \infty$ as A's spaceship approaches the horizon, $r \to r_g$. The proper time, which is the time measured by Alice herself,

$$\tau = -\int_{r_0}^{r} \frac{e\sqrt{r}dr}{\sqrt{\left[r_g - r(e^2 - 1)\right]}} < \infty \tag{12}$$

turns out to be finite. This illustrates a manifestation of the most dramatic time delay: for distant observers (but actually for all observers exterior to the horizon), Alice's frame of reference would need an infinite time to reach the event horizon, while a finite time elapses for the co-moving observer, Alice herself. Another aspect of this outcome has already been mentioned. The speed, V, of the freely falling test particle as measured by a static observer, SO, follows from the expression (see also [5]),

$$u_{SO}U = f\dot{t}\frac{1}{\sqrt{f(r)}} = \frac{1}{\sqrt{1-V^2}} \tag{13}$$

One finds then a general outcome: the speed of a test particle radially freely falling

$$V^2 = \frac{e^2 - f(r)}{e^2} \xrightarrow[f \to 0]{} 1 \tag{14}$$

approaching the event horizon tends to the value of the speed of light in the vacuum. And this result is independent of the initial conditions. One may ask: how would that speed be changing inside the horizon? We discuss this question subsequently.

3.2. Generalized Doppler shift: how to fix the instant of crossing of the Schwarzschild BH horizon

It is a well-known fact that due to the equivalence principle, an observer confined within a frame freely falling towards the horizon cannot identify the instant at which he/she crosses the horizon and if a black hole is large enough, such an observer would harmlessly cross the horizon without even noticing [6]. On the other hand, one can quite precisely determine that instant. How is this seeming contradiction possible?

Before resolving this, let us recollect a well-known result, that of the gravitational frequency shift. In order to do this, one considers radial signals of a fixed frequency, $\overline{\omega}$ emitted at r_0 (the location of the Mother Station) and recorded by a static observer at $r > r_g$. The wave vector k of those radial light rays, $k = (k^t, , k^r, 0, 0)$ is (see Eqs. (6), (7)):

$$k^t = \frac{\omega}{f}, \tag{15}$$

$$k^r = \pm\omega, \tag{16}$$

where $\pm$ corresponds to out- and ingoing rays, respectively. If MS emits such a signal with frequency

$$\omega^e_{MS} = \frac{\omega}{\sqrt{f(r_0)}} \equiv \overline{\omega}, \tag{17}$$

SO records it at r and measures its frequency as

$$\omega_{SO} = u_{SO}k = \frac{\omega}{\sqrt{f(r)}} \tag{18}$$

so

$$\frac{\omega^r_{SO}}{\omega^e_{MS}} = \sqrt{\frac{f(r_0)}{f(r)}} \xrightarrow[f \to 0]{} \infty. \tag{19}$$

The frequency recorded by SO is indefinitely blueshifted: when r tends to r_g, $f \to 0$.

When such radial signals are recorded by Alice, $\omega_A = u_A k$, at her instantaneous position at r, then she finds (see Eqs. (6)–(9) and (15, 16)

$$\omega^r_A = \frac{\frac{\omega}{\sqrt{f(r_0)}}}{1 + V} \tag{20}$$

where V is the speed of her spaceship as measured by SO (placed at r) (see Eq. (14)).

Exchanging such signals, one can observe a (generalized) Doppler shift of the following form [7]:

$$\frac{\omega^r_A}{\omega^e_{MS}} = \frac{1}{1 + \sqrt{\frac{e^2 - f(r)}{e^2}}} = \frac{1}{1 + V} \tag{21}$$

and

$$\frac{\omega^r_{MS}}{\omega^e_A} = 1 - V \tag{22}$$

The meaning of result (22) is as follows: signals coming from a frame infalling towards the black hole horizon are indefinitely redshifted (and ultimately disappear from the screens/ sensors)—such a journey seems to take infinitely long for external observers. This confirms our former conclusion. The result (21) on the other hand means that the Doppler shift of signals coming from MS allows Alice to identify the horizon quite precisely—the Doppler shift reaches a value of ½ on the horizon.

3.3. Image collision or the 'touching ghosts' anomaly

With the speed of free fall tending to the speed of light in a vacuum, the generalized Doppler shift as characterized by Eqs. (21) and (22) and the dramatic form of the time delay in this case, this leads to yet another anomaly—*image collision* [8] or *touching ghosts* [9]. Signals emitted by Alice located within the infalling frame appear to get "frozen" in the proximity of the horizon (see, however, [10, 11]).

Let us consider another observer, B (Bob), whose spaceship also starts from MS, following Alice's spaceship. Alice and Bob exchange electromagnetic signals; how (when) does Bob perceive the instant of Alice's crossing of the horizon? The answer has been referred to as 'image collision' or 'touching ghosts' and it is as follows [8, 9]. Alice sends an encoded message: a signal that means 'I am crossing the horizon' (at the instant when her Doppler shift is half); Bob receives that message at the instant when he himself crosses the horizon.

An interesting fact is that this effect, originally illustrated by means of Kruskal-Szekeres coordinates, may be interpreted in a general manner, without reference to any specific system of coordinates. Indeed, if Bob received such a message before crossing the horizon, that information would be transmitted to our part of the universe; this would contradict the fact that the horizon crossing can never be observed.

3.4. Photon sphere

In the case of null geodesics in the equatorial plane, the wave vector components are as follows:

$$k^t = \frac{\omega}{f}, \quad k^\varphi = \frac{l}{r^2} \tag{23}$$

$$k^r = \pm\sqrt{\omega^2 - f\frac{l^2}{r^2}} \equiv \pm l\sqrt{\frac{1}{b^2} - V_{eff}(r)} \tag{24}$$

where b is a so-called impact parameter. The function $V_{eff}(r) = f\frac{r^2}{r^2}$ is regarded as an 'effective potential' for null geodesics (see **Figure 1**). The shape of null geodesics depends on the value of b. The deflection angle

$$\int d\varphi = \pm\int \frac{dr}{r^2\sqrt{\frac{1}{b^2} - V_{eff}(r)}}$$

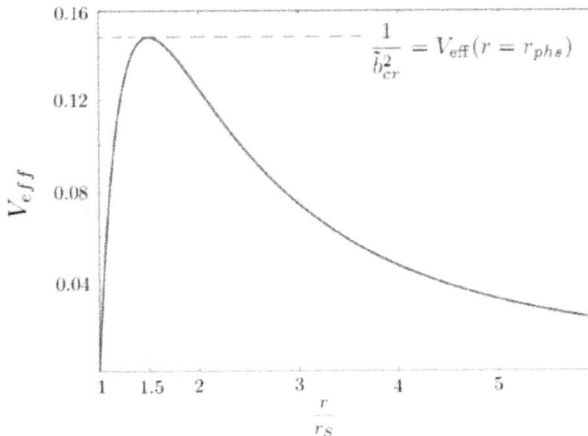

Figure 1. Effective potential $V_{eff}(r) = f\frac{r^2}{r^2}$ in the case of Schwarzschild spacetime (horizontal axis—r expressed in units, $r_S \equiv r_g = 2M$).

is small for large values of b—light rays are only slightly deflected. It grows indefinitely as the impact parameter value tends to its critical value, $b_{cr}^{-1} = V_{eff}(r_{phs})$. The impact parameter b_{cr} corresponds to the so-called 'photon sphere' composed of circular trajectories, $r = r_{phs}$,

$$r_{phs} = \frac{3}{2}r_g \equiv 3M \tag{25}$$

which are (unstable) null geodesics:

$$\left(k^t, 0, 0, , k^\varphi\right) = \left(\frac{\omega}{f(r_{phs})}, 0, 0, \frac{l}{r_{phs}^2}\right) \tag{26}$$

3.5. The shape of light cones

It should be noted that in approaching an event horizon, the shape of a light cone evolves in a characteristic manner. Indeed, observing radial in- and outgoing signals

$$ds^2 = f dt^2 - \frac{1}{f}dr^2 = 0 \tag{27}$$

one finds,

$$\frac{dr}{dt} = \pm f \xrightarrow[r \to r_g]{} 0 \tag{28}$$

which may be illustrated as a sequence of vanishingly narrow cones.

4. Interior of Schwarzschild BH

In order to describe the interior of the horizon of the Schwarzschild spacetime, one can follow an approach proposed by Doran et al. [12]. These authors showed that discussing the problem of an empty, but dynamically changing spacetime, one finds, using specific boundary conditions, the metric (4) of the interior of the Schwarzschild spacetime, that is,

$$ds^2 = \left(1 - \frac{2M}{r}\right)dt^2 - \frac{1}{\left(1 - \frac{2M}{r}\right)}dr^2 - r^2 d\Omega^2 \tag{29}$$

for $r < r_g = 2M$ (see also [13]). This means that formally one can use Schwarzschild coordinates also for the interior of the horizon, but then one must remember about the exchange of the roles of the t and r coordinates. Inside the horizon, r plays the role of a temporal coordinate: it changes from r_g to 0 and $dr < 0$; t plays the role of a spatial coordinate, changing between $-\infty$ and $+\infty$ with dt taking both positive and negative values. The important consequence is a change of the symmetry of the system: instead of a static, spherically symmetric spacetime, one encounters a homogeneous, spherically symmetric and dynamically changing spacetime; energy is no longer conserved but (due to the homogeneity along the t-axis), appropriately, the t-momentum component is conserved.

Therefore, one can consider spacetime (29) as representing the interior of a Schwarzschild black hole. Accordingly, analogues of the phenomena described above outside the horizon will be analyzed.

First, one introduces a class of resting observers, RO, that is, those, whose spatial coordinates, t, θ, φ are fixed. Then, the velocity u_{RO} four vector's only nonvanishing component is a temporal one,

$$u_{RO} = -\sqrt{-f}\, \partial_r. \tag{30}$$

The class of infalling test particles located in Alice's frame of reference is described in the same way as given outside the horizon (Eqs. (6)–(9))—in this case, however, $r < r_g = 2M$, so $f < 0$. In this region, ingoing (−) and outgoing (+) null geodesics (that are planar) described as

$$f\frac{dt}{d\sigma} = \pm\omega \qquad r^2\frac{d\varphi}{d\sigma} = l \qquad \frac{dr}{d\sigma} = -\sqrt{\omega^2 - f\left(\frac{l^2}{r^2} + \eta\right)} \tag{31}$$

differ from their counterparts outside the horizon by a small but important feature—the $\pm$ sign is designated to a spatial coordinate, namely the r-coordinate outside the horizon and the t-coordinate inside the horizon. Having said this, one may now discuss specific effects (see [14, 15]).

4.1. The speed of an infalling test particle

A test particle located in A's framework (Eqs. (6)–(9)), $l = 0$, is freely falling FF. Then, a resting observer (30) measures its (squared) speed $\tilde{V}^2$ as follows:

$$u_{RO}U_{FF} = -\frac{1}{f}\sqrt{-f}\sqrt{e^2 - f} = \frac{1}{1 - \tilde{V}^2}. \tag{32}$$

One finds then that (c.f. Eq. (14))

$$\tilde{V}^2 = \frac{e^2}{e^2 - f}. \tag{33}$$

This is, at first sight, a rather unexpected outcome: the speed is given by a formula inverse to the one obtained outside the horizon, Eq. (14). Another aspect of this result is revealed when one illustrates the speed outside and inside the horizon as measured by observers that are static, SO, and resting, RO (30), respectively (see **Figure 2**).

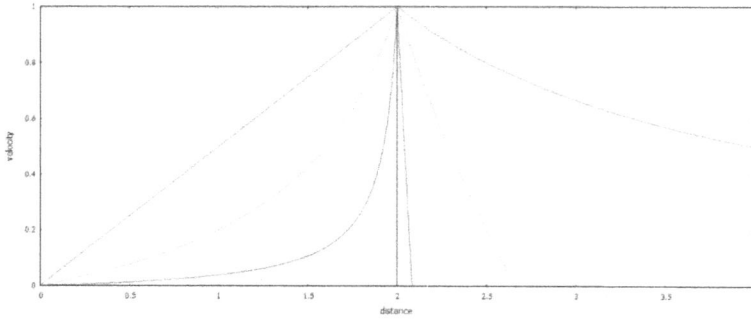

Figure 2. Values of 'velocity' V^2 measured by SO (outside horizon) and $\tilde{V}^2$ by RO (inside horizon) of different test particles in the Schwarzschild spacetime. The red curve corresponds to $e = 1$, $\tilde{V}^2 = \frac{r}{r_g}$, the green curve to, $e = 0.5$ and the blue one to $e = 0.2$. The vertical line represents the horizon located at $r_g = 2$ (horizontal axis — r expressed in units M).

4.2. The Doppler shift

Let us consider an analogy of the generalized Doppler effect inside the horizon.

4.2.1. Frequency shift of signals coming from MS

4.2.1.1. Resting observers

One can start from an analogy of the gravitational frequency shift: a resting observer (30) records radially incoming signals coming from the Mother Station. Then, according to Eq. (18), the frequency shift is

$$\frac{\omega^r_{RO}}{\omega^s_{MS}} = \sqrt{\frac{f(r_0)}{-f(r)}} \longrightarrow \begin{cases} \infty & r \to r_g \\ 0 & r \to 0 \end{cases} \tag{34}$$

One finds then that the gravitational frequency shift of the signals coming from MS and recorded by static, SO, and resting, RO, observers, outside and inside the horizon, respectively, as having a symmetric form with respect to the horizon itself (see Eqs. (19) and (34)).

4.2.1.2. Freely falling observers

The frequency shift of signals coming from MS and recorded by Alice, who is radially freely falling, is

$$\frac{\omega^r_A}{\omega^s_{MS}} = \frac{1}{1 + \sqrt{\frac{e^2-f}{e^2}}} \longrightarrow \begin{cases} \frac{1}{2} & r \to r_g \\ 0 & r \to 0 \end{cases} . \tag{35}$$

Expression (35) is the same as its counterpart outside the horizon (21): it turns out that the frequency shift is a continuous and decreasing function from 1 to 0 during the trip through the horizon; as emphasized earlier, with the factor $\frac{1}{2}$ marking the horizon (see **Figure 3**)

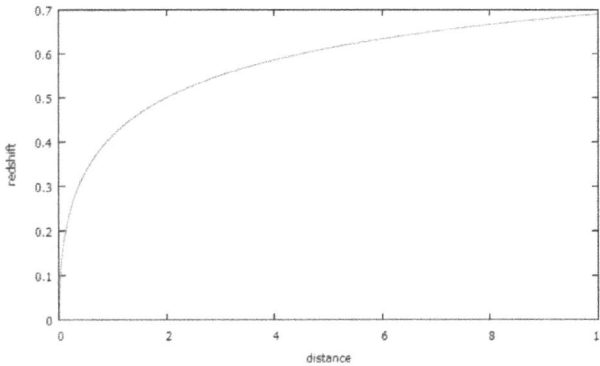

Figure 3. Monotonic and continuous change of the frequency ratio $\frac{\omega_A^r}{\omega_{MS}^r}$ (redshift—Vertical axis) outside and inside the horizon (horizontal axis—r expressed in units M, $r_g = 2$).

4.2.2. Frequency shift of signals inside the horizon of BH

One can consider the exchange of signals by observers at rest inside the horizon. One can distinguish two types of signals: going along the direction of homogeneity, that is, the t-axis, and signals propagating perpendicularly to this axis.

4.2.2.1. Signals propagating along the t-axis

The frequency shift of signals exchanged by two observers at rest at t_1 and t_2 depends on the emission instant, r_1 (recording instant r_2 is fixed by the distant t_1, t_2):

$$\frac{\omega^r(t_2)}{\omega^e(t_1)} = \frac{\sqrt{-f(r_1)}}{\sqrt{-f(r_2)}} \xrightarrow{r_2 \to 0} 0. \tag{36}$$

One finds that in this case, the frequency redshift tends to zero at the ultimate singularity (see **Figure 4**).

4.2.2.2. Signals propagating perpendicularly to the t-axis

The wave vector of signals propagating perpendicularly to the t-axis has two non-vanishing components $k = k^r \partial_r + k^\varphi \partial_\varphi$ (because of the planar character of the trajectory, one can choose $\theta = \frac{\pi}{2}$, i.e. the equatorial plane). Then, the frequency shift, for two static observers placed within this plane perpendicular to the t-axis, is given by

$$\frac{\omega^r(\varphi_2)}{\omega^e(\varphi_1)} = \frac{r_1}{r_2} \xrightarrow{r_2 \to 0} \infty. \tag{37}$$

One finds an indefinite blueshift at the ultimate singularity (see **Figure 5**).

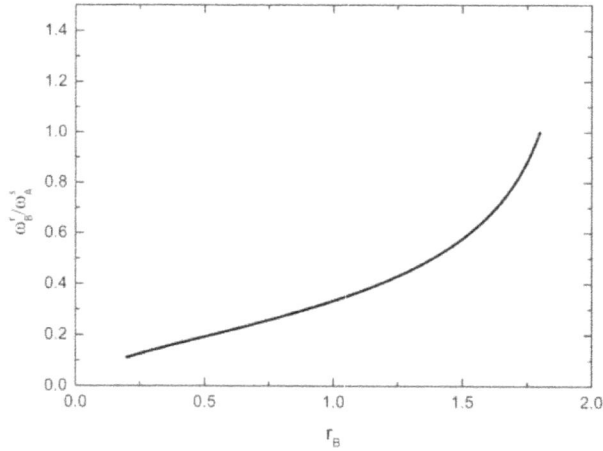

Figure 4. Frequency redshift (vertical axis) for the signal propagating along homogeneity direction between instants, $r_B = 0.9r_g = 1.8$ and $r_A = 0.2r_g = 0.4$ as a function of r (horizontal axis—In M units, $r_g = 2$).

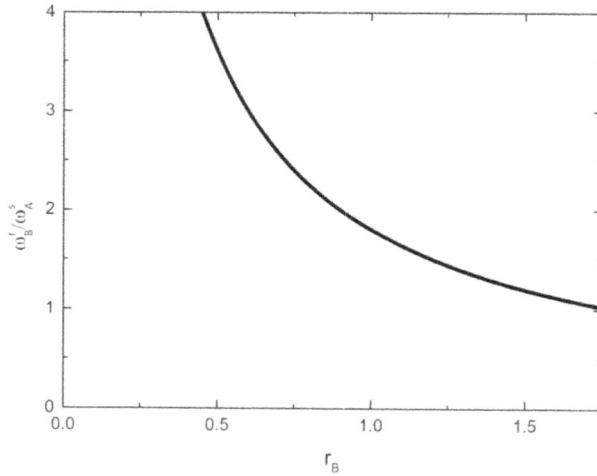

Figure 5. Frequency blueshift (vertical axis) for the signal propagating perpendicularly to the homogeneity direction, between instants, $r_B = 0.9r_g = 1.8$ and $r_A = 0.3r_g = 0.6$ as a function of r (horizontal axis—In units M, $r_g = 2$).

4.3. Photon sphere analogue

Null geodesics propagating perpendicularly to the t-axis resemble trajectories belonging to the photon sphere. Indeed, they are determined by the wave vector,

$$(0, k^r, 0, , k^\varphi) \equiv \left(0, -\sqrt{\frac{-1}{f}\frac{l^2}{r^2}}, 0, \frac{l}{r^2}\right) \tag{38}$$

having only one spatial component, the angular component, k^φ corresponding to a circular-like motion. There is one significant feature distinguishing these null geodesics from the circular trajectories of radius $r = r_{phs}$ outside the horizon: they circulate over a sphere of an ever-decreasing radius. One can see from the null condition:

$$-\frac{1}{f}dr^2 - r^2 d\varphi^2 = 0 \tag{39}$$

that the rate of change of the radius of such a sphere is proportional to r, which is a temporal-like (decreasing) coordinate. Therefore, one finds inside a black hole an interesting phenomenon: a photon sphere analogue. Outside the horizon, a light ray belonging to the photon sphere can (in principle, as it is a circular trajectory of unstable equilibrium) unwind infinitely many times. One can ask then: inside a black hole, how many times can a light ray orbit along a photon sphere analogue before reaching the ultimate singularity?

The answer to this question is quite unexpected: it is exactly *a single half rotation*.

Indeed, by using Eq. (39), one obtains

$$\Delta\varphi = \int \frac{dr}{r\sqrt{\frac{-f}{r^2}}} = \int_0^{2M} \frac{dr}{\sqrt{r(2M - r)}} = \pi \tag{40}$$

This means that the angle traversed by a light ray is equal in this case to π. A general property is that the deflection angle for a light ray within the BH horizon cannot exceed π.

5. The horizon of a Schwarzschild BH

Among various interesting properties of the Schwarzschild BHs horizon, there are at least two that are relevant to our discussion.

The first relates to the speed of an object crossing the horizon. As described earlier, the value of the speed of Alice's spaceship tends to the value of the speed of light c as it approaches the horizon. Does that speed take the value c on the horizon? There are no observers residing on the horizon, but other observers, crossing the horizon, would in principle be able to perform such a measurement. Performing this kind of thought experiment, one obtains the following: the speed of Alice's spaceship crossing the BH horizon is less than the speed of light. The value of that speed depends on the initial conditions.

The second is linked to any outgoing light ray trapped at the horizon. It may be a signal emitted by Alice at the instant she was crossing the horizon with the encoded message: 'I am crossing the horizon now'. If it was a signal of some specific frequency, what would be its frequency as recorded by Bob, when he crossed the horizon? It turns out that such a signal 'ages': it is redshifted and the value of the redshift becomes greater as the original distance between Alice and Bob increases [15].

6. The meaning of a strong gravitational field

Let us underline the rather unexpected and counterintuitive observations that accompany the presence of the event horizon of a Schwarzschild BH. The strange intimate symmetry of the outer versus inner region: static observers outside the horizon and observers at rest inside the horizon measuring the Doppler shift of signals incoming from MS would record basically the same outcomes. The speed of a test particle falling towards the BH appears to be impeded after crossing the horizon. As described elsewhere, the speed of a test particle uniformly accelerated inside the horizon after reaching its maximal value starts to diminish. A null geodesic follows exactly half a circular orbit within the horizon. Signals exchanged within the horizon seem to mimic the cosmological model expanding along one specific direction and contracting perpendicularly to this direction. All of these are manifestations of the presence of such a strong gravitational field that the event horizon of the BH is developed.

Inside the horizon of a Schwarzschild BH, one comes across a unique phenomenon: an interchange of the roles of the r and t coordinates. Outside the horizon $r > r_S$, the radial coordinate is an ordinary spatial coordinate, which may change from r_S to ∞ in both directions, $dr = \pm|dr|$ and the time coordinate t is a temporal one, that is, such that, $dt > 0$. Inside the horizon $r < r_S$, and coordinate r becomes a temporal one: r changes from r_S to 0 and $dr < 0$; coordinate t then plays the role of a spatial coordinate: $-\infty < t < \infty$ and $dt = \pm|dt|$.

Such an interchange results in a dramatic difference of the symmetry properties of the spacetime. As mentioned earlier, the Schwarzschild spacetime outside the horizon is static, independent of time and isotropic; this results in the conservation of energy and angular momentum, respectively. Inside the horizon, the spacetime is still independent of t but this is now a spatial coordinate in that spacetime, leading to t-component momentum conservation; it is no longer static but instead dynamically changing, being r-dependent. Inside the horizon of the Schwarzschild BH, spacetime is cylindrical-like, homogeneous along the t-axis and spherical-like, of radius r perpendicularly to this axis (see also [2, 16]).

All of this presents the above-seemingly unexpected or counterintuitive phenomena in a new perspective. The speed of the infalling test particle is measured as 'distance'/'time' so the interchange of the roles of 'distance' and 'time' leads to the inverse expressions to those exterior to the horizon, V and interior, $\tilde{V}$; hence, the speed turns out to decrease inside the horizon. The cylindrical-shape BH interior is a dynamically changing spacetime: expanding along the t-axis and contracting perpendicularly to this axis. This results in both red- and blueshifts, respectively [12, 17]. Hence, it actually *is* a realization of a specific *cosmology*. The fact that light rays propagating perpendicularly to the cylinder axis occupy a semicircular photon sphere analogue is found to have a deeper significance [18]. The same value π is found for other kinds of black holes, and this appears to be a fundamental discovery; it may be a symmetry property linked to a 'new physics' of black holes [19]. Also, other observations may need deeper analysis but, whatever the interpretation, they are caused by the strong gravitational fields that form the BH horizon.

Let us emphasize that the common sense property of the BH, namely. 'nothing, not even light can leave their interior' takes on a new sense now: crossing the event horizon, a test object can never reach it again as this would mean travelling backwards in time.

There is a more formal interpretation of the interchange of the role of radial and temporal coordinates in the theory of relativity. The Killing vector representing time independence symmetry, being time-like outside the horizon becomes space-like inside the horizon—this actually means that the time-like component of the momentum four vector is converted into a space-like momentum component, respectively. This opens the door for radiation emitted by black holes—Hawking radiation.

7. Astrophysical black holes

Generations of thermonuclear reactions support stars against gravitational collapse [3, 20]. The first stage is a process of hydrogen burning to make helium. When a substantial amount of hydrogen is exhausted, gravitational contraction raises the temperature until helium burning, the so-called triple alpha process, can start. This evolution eventually leads, for massive stars, to the last stage where an element with the largest binding energy per nucleon, $^{56}_{26}Fe$, is produced. What happens then?

One can consider the state of a star of mass M and radius R, which exhausted its thermonuclear fuel, $T = 0$. It is supported by a nonthermal pressure, due to the fermionic nature of electrons, protons and neutrons. There are two competing contributions to the energy of such an object. A negative one arises from a gravitational origin

$$E_g \propto -\frac{M^2}{R} \qquad (41)$$

and a positive one, the kinetic energy of the electronic gas:

$$E_k \propto nR^3 \langle E \rangle \qquad (42)$$

where n denotes the density of electrons and $\langle E \rangle$ is the electronic mean energy. Taking the following relation between the characteristic electron momentum, p_F, and the corresponding wavelength, $\lambda \propto n^{-1/3}$,

$$p_F \propto \lambda^{-1} \propto n^{1/3} \qquad (43)$$

one obtains for a nonrelativistic range of energies, $\langle E \rangle \propto p_F^2$

$$E_k \propto \frac{M^{5/3}}{R^2}.$$ (44)

It appears that the kinetic energy term dominates in the range of decreasing values of R, preventing further contraction. However, for more massive stars, higher energies are available and the electrons would be regarded as relativistic, $\langle E \rangle \propto p_F$ and then,

$$E_k \propto \frac{M^{4/3}}{R}$$ (45)

In such a case for a mass M larger than the Chandrasekhar limit, $M_{WD} \approx 1.4 M_\odot$ (for white dwarfs), the pressure of the electron gas could not support a star against its gravitational contraction.

For even more massive stars, one comes across inverse beta decay leading to the formation of a neutron star core. In such a case, the Pauli exclusion principle, this time for neutrons, prevents gravitational collapse, up to some specific limit, $M_{cr} \propto 2 - 3\ M_\odot$. For masses larger than this limiting case, nothing can stop the ongoing gravitational collapse eventually leading to a singular state of matter—a black hole.

8. Entropy and Hawking radiation

In early 1970s, it was indicated by Bekenstein [21] and Hawking [22] that BH entropy is proportional to their surface area:

$$S = k_B \frac{4\pi r_S^2}{4 l_P^2} = k_B \frac{4\pi M^2}{l_P^2}$$ (46)

where l_P denotes the Planck length and k_B Boltzmann's constant. It was also recognized that BHs may be regarded as simple thermodynamic systems (the black hole 'no hair' theorem) characterized by three parameters, mass M, angular momentum J and charge Q. Accordingly, one can identify four different kinds of black hole: Schwarzschild (nonrotating and uncharged, characterized by their mass M), Reissner-Nordstrom (charged but nonrotating, characterized by M and Q), Kerr (rotating, characterized by M and J) and Kerr-Newman (rotating and charged, characterized by M, J and Q). In the case of Schwarzschild spacetime, one can apply a simple thermodynamic formula [21, 22].

$$dU = TdS$$ (47)

and identifying $U = M$ to determine the BH temperature T_{BH} as being proportional to its inverse mass,

$$T_{BH} = \frac{\hbar c^3}{8\pi M k_B G}$$ (48)

where we use standard notation. It was Hawking's idea that black holes should lead to a new kind of uncertainty [23], other than the one having a quantum mechanical origin. When matter or radiation falling in towards a black hole crosses its horizon, the information it carries is inevitably lost. This led to two controversies. Firstly, information itself is lost. Secondly, one can consider black hole formation due to the gravitational collapse of matter (or radiation) as the unitary evolution of a pure quantum state. After the formation of the horizon, further evolution has to be regarded in terms of mixed states due to the loss of information. This means the breakdown of quantum mechanical predictability. Both elements of such an information problem, loss of information and breakdown of unitary quantum evolution, were objected to from the very beginning.

Hawking himself [24] formulated the idea of black hole decay. Due to the existence of an event horizon and the conversion of one of the Killing vectors from a temporal to a spatial one, a pair of entangled particles, one of positive and one of negative energy, would be created in the proximity of the horizon. Two scenarios are then possible when one (the one with negative energy) or both of the particles fall behind the horizon. The point is that the particle with negative energy could not 'survive' in our part of the Universe for fundamental reasons, but it could exist within the horizon. This is so because the energy, the t-component of the particle momentum vector within the horizon, takes on a spatial character, so it might then be either positive or negative. Hence, one of the particles, the one with positive energy, departs to infinity, being recorded as Hawking radiation and the other member of the pair, with negative energy, falls behind, 'tunnelling through' [25] the horizon and reducing the BH mass. This is the meaning of BH evaporation. Hawking evaporation is the radiation of a black body of temperature, T_{BH} (8.2).

Therefore, BHs turn out to be evaporating nonequilibrium systems with a decay time

$$t_{BH} \cong 10^{74} \left(\frac{M}{M_\odot} \right)^3 \qquad (49)$$

fifty seven orders of magnitude larger than the age of the Universe for moderate BH masses M. According to the generalized second law of thermodynamics, the entropy during evaporation is an increasing function of time. Indeed, during evaporation, the BH entropy decreases,

$$dS_{BH} = -\frac{dU}{T_{BH}} \qquad (50)$$

yet the entropy of the respective BH radiation is larger by one-third [26].

$$dS_r = \frac{4}{3} \frac{dU}{T_{BH}}. \qquad (51)$$

One may suspect that information lost due to the presence of the horizon may be retrieved due to evaporation, thus restoring this fundamental aspect of quantum mechanical unitary evolution [27–29]. A closer scrutiny shows that this is not so obvious: at the initial stages of the BH decay, both BH and radiation are close to their maximally mixed states, thus no information is retrieved. Although the process of releasing information might be of a non-perturbative character, the information problem (referred to as the information paradox) still remains unsolved. It was indicated that smooth quantum mechanical unitary evolution should lead to the breakdown of the smoothness of the proximity of the event horizon, leading to a 'firewall' [30]. This concept was objected to in more recent papers [31–33]; nevertheless, the information paradox is still far from being removed. It may currently be formulated in many different ways and one of those ways can be expressed as follows:

Hawking radiation consists of particles born as entangled pairs; those recorded far away are then entangled with a diminishing BH. Finally, the BH disappears. What, then, are those particles recorded at distant locations still entangled with [34]?

9. Final remarks

The purpose of this presentation was to illustrate selected features of strong gravitational fields. Black holes are the sources of the strongest gravitational field in the sense that an event horizon has developed. Let us briefly consider the point 'black but not dark'. The presence of black holes may be recognized primarily due to gravitational interactions: the dynamics of their environment. In this sense they may be regarded as a component of a dark matter sector. Accepting such an oversimplified or naïve point of view for a while, one may ask about the character of this component. Partly, the answer is obvious: this is baryonic matter, as massive stars collapsing into black holes are composed of baryonic matter. But due to instability, there are no extremely massive stars, so BHs of millions of solar masses have a different origin (eternal black holes), so they might not necessarily be composed of baryonic matter. In principle, as they evaporate, they emit radiation; also, they could be charged so they could therefore affect their environment not only gravitationally. Hence, although they are black they are hardly 'dark'. As mentioned at the beginning of this exposition, BHs constitute a small fraction of the density of baryonic matter, so they are interesting objects in the Universe rather for the local properties imposed by their gravitational field, than for other reasons (at least so it seems to us at the moment).

The outcome of the presence of the horizon of the BH is a dramatic difference in the symmetry properties of the exterior and interior of the BH. Energy conservation related to the time-like Killing vector is changed into a corresponding momentum component conservation as the Killing vector is converted into a space-like one. That is a consequence of the static spacetime outside the horizon being transformed into a homogeneous one, along the t-direction, but it also becomes a dynamically changing spacetime inside the horizon: expanding along the homogeneity direction and contracting perpendicularly to that direction. On the one hand, this leads to the information paradox. On the other hand, the presence of the BH's event horizon may be interpreted as an interchange of the roles of the time and radial coordinates. And this leads to unexpected scenarios, with some surprising processes and phenomena taking place outside the horizon yet with even more striking properties of the interior of the horizon. It should be underlined that the discussion presented here has dealt mostly with eternal BHs, which have not been created due to gravitational collapse but rather have existed forever (since the Big Bang). However different these may seem, they have a lot in common. They both decay due to Hawking radiation [2]; as suggested by various authors [16], the interior of gravitationally collapsing black holes is also of a cylindrical shape, and both eternal and collapsing BHs share one more common but bizarre property, their volume is *infinite* [16, 35]. Hence, though it is not guaranteed that the interiors are the same their properties might turn out to be quite similar. But there is a still a deeper problem of a much more fundamental character: could the interior of black holes be described by the approach presented here? Or more specifically, could a very strong gravitational field, inside the BH horizon, be described in terms of the theory of relativity? Or is a new physical approach necessary, as emphasized by G. t'Hooft [19] (see also [36]) involving quantum mechanical aspects also? As usual, the answer will come in time, but even if the answer is satisfactory, in this case, it will probably never be the final word.

Author details

Andrzej Radosz[1]*, Andy T. Augousti[2] and Pawel Gusin[1]

*Address all correspondence to: andrzej.radosz@pwr.edu.pl

1 Wroclaw University of Science and Technology, Wroclaw, Poland

2 Kingston University London, Kingston, UK

References

[1] Weinberg S. Gravitation and Cosmology: Principles and Applications of the General Theory of Relativity. New York: Wiley; 1972

[2] Frolov VP, Novikov' ID. Black Hole Physics: Basic Concepts and New Developments. Dordrecht, Netherlands: Kluwer Academic; 1998

[3] Hartle JB. Gravity: An Introduction to Einstein's General Relativity. Reading, MA: Addison-Wesley; 2003

[4] Landau L, Lifshitz E. The Classical Theory of Fields. 3rd. ed. Reading, Mass: Addison-Wesley; 1977

[5] Janis A. Motion in the Schwarzschild field. Physical Review D. 1977;**15**:3068

[6] Misner CW, Thorne KS, Wheeler JA. Gravitation. New York: Freeman; 1973

[7] Radosz A et al. The Doppler shift in Schwarzschild spacetime. Physics Letters A. 2009;**801**: 373

[8]   Müller T. Falling into Schwarzschild black hole. General Relativity and Gravitation. 2008; **40**:2185-2199

[9]   Augousti A et al. Touching ghosts: Observing free fall from an infalling frame of reference into a Schwarzschild black hole. European Journal of Physics. 2012;**33**(1)

[10] Kassner K. Why ghosts don't touch: A tale of two adventurers falling one after another into a black hole. European Journal of Physics. 2016;**38**(1)

[11] Toporensky AV, Zaslavskii OB. Redshift of a photon emitted along the black hole horizon. European Physical Journal C: Particles and Fields. 2017;**77**:179

[12] Doran R, Lobo FSN, Crawford P. Interior of a Schwarzschild black hole revisited, [arXiv:gr-qc/0609042v2]

[13] Hamilton AJS, Lisle JP. The river model of black holes. American Journal of Physics. 2008; **76**:519-532

[14] Crawford P, Tereno I. Generalized observers and velocity measurements in General Relativity, [arXiv:gr-qc/0111073]

[15] Toporensky AV, Zaslavskii OB, Popov SB. Unified approach to redshift in cosmological / black hole spacetimes and synchronous frame arXiv:1704.08308v2 [Gr-Qc]; 2017

[16] Christodoulou M, Rovelli C. How big is a black hole? Physical Review D. 2015;**91**:064046

[17] Radosz A, Gusin P, Formalik F, Augousti A. 2018. (in preparation)

[18] Sanchez N, Whiting BF. Quantum field theory and the antipodal identification. Nuclear Physics B. 1987;**283**:605

[19] G. 't Hooft, The firewall transformation for black holes and some of its implications, arXiv: 1612.08640v3 [gr-qc] (2017)

[20] P.K. Townsend, Black Holes, arXiv:gr-qc/9707012v1 4 Jul 1997

[21] Bekenstein JD. Black holes and entropy. Physical Review D;**7**:233

[22] Hawking SW. Black holes and thermodynamics. Physical Review. 1976;**D13**:191

[23] Hawking SW. Breakdown of predictability in gravitational collapse. Physical Review D. 1976;**14**:2460

[24] Hawking SW. Black hole explosions? Nature. 1974;**248**:30

[25] Parikh MK, Wilczek F. Hawking radiation as tunneling. Physical Review Letters. 2000;**85**: 5042

[26] Zurek WH. Entropy evaporated by a black hole. Physical Review Letters. 1982;**49**:1683

[27] Page DN. Comment on "Entropy evaporated by a black hole. Physical Review Letters. 1983;**50**:1013

[28] Page DN. Information in black hole radiation. Physical Review Letters. 1993;**71**:3743

[29] Braunstein SL, Pirandola S, Życzkowski K. Better Late than Never: Information Retrieval from Black Holes. Physical Review Letters. 2013;**110** 101301

[30] Almheiri A, Marolf D, Polchinski J, Sully J. Black holes: Complementarity or firewalls? JHEP. 2013;**2013**:62 [arXiv:1207.3123]

[31] Lowe DA, Thorlacius L. Pure states and black hole complementarity. Physical Review D. 2013;**88**:044012

[32] Banks T, Fischler W. Holographic space-time does not predict firewalls, [arXiv:1208.4757

[33] Mathur S. A model with no firewall. [arXiv:1506.04342]

[34] Radosz A, Gusin P, Roszak K. Disentanglement and black holes: Information problem. Acta Physica Polonica A. 2017;**132**:132

[35] Gusin P, Radosz A. The volume of the black holes - the constant curvature slicing of the spherically symmetric spacetime. Modern Physics Letters A. 2017;**32**:1750115

[36] Mathur S. Black holes and beyond, arXiv:1205.0776v1 [hep-th] 3 May 2012

www.ingramcontent.com/pod-product-compliance
Lightning Source LLC
Chambersburg PA
CBHW081244190326
41458CB00016B/5914